SUSTAINABILITY IN THE FOOD INDUSTRY

SUSTAINABILITY IN THE FOOD INDUSTRY

Cristina García Jaime

Delve PUBLISHING

www.delvepublishing.com

Sustainability in the Food Industry

Cristina García Jaime

Delve Publishing

2010 Winston Park Drive,

2nd Floor

Oakville, ON L6H 5R7

Canada

www.delvepublishing.com

Tel: 001-289-291-7705
001-905-616-2116

Fax: 001-289-291-7601

Email: orders@arclereducation.com

ISBN: 978-1-77361-400-7 (Hardcover)

ABOUT THE AUTHOR

Cristina Garcia Jaime is a Spanish food specialist who studied Food Quality and Technology in the University of Granada and Human Nutrition and Dietetics in Pablo de Olavide University (Seville). She has worked in different food areas: scientific, industry and clinic. Currently, she has founded a company established in Europe and called "Foodise" dedicated to help small business and companies to manage all issues related to food.

TABLE OF CONTENTS

LIST OF FIGURES

LIST OF TABLES

PREFACE

Modern food systems are undergoing a technical and structural transformation amid increasing globalization and preference for sustainability. The main aim of this book is to give the reader a complete understanding of sustenance in the food industry and act as a reference guide to the overall technique and activities that are involved in preserving, processing, and manufacturing food products. It also reflects the additional qualitative efforts required to compare the results that already exist and those that are under development in the industry with the aim of developing commercial logic for sustainability.

Various chapters of the book are arranged in a logical, sequential way, starting with a brief introduction on sustainability in the food sector, followed by an overview of the potential role and challenges in the food industry and several principles of sustainability for food products. After understanding the basic concepts of the food sector, further focus is given on the quality, safety, and sustainability in the food manufacturing. In this section, the role or HACCP system, risk management using HACCP system, and importance of this system for a sustainable development are discussed. This book also contains information about the entire food distribution and supply chain system, distribution network design, and packaging and labeling of the food products. In the next part, the book takes readers' attention towards several sustainability strategies, the issues that are faced in sustainability and the top trends and sustainable practices in the food industries.

"*Sustainability in the food industry*" book also provides a brief overview on sustainability as innovation and the green growth of sustainability in the food industries. Outlining of unique challenges like issues of logistics, new supply routes, increasing dynamics and investment needs, etc. are discussed in detail. Strategies and methods that are taken into consideration to improve food supply chain for enhancing the sustainability and resilience, form a significant part of the content. Corporate Social Responsibility (CSR) in the supply chain is one of the most important topic which is separately discussed. This section is dedicated towards the importance of CSR in the food supply chain, as this industry is fully aware of its social responsibility and is committed to making a positive contribution to society and the environment.

At the end, the book covers the recent developments and advancement in the food sectors, the important drivers of change affecting the food system, and the emerging global trends in the food sector. Also, the next generation technologies in the food systems are discussed in detail.

Overall this book will help and guide the reader to completely understand the foundations of the food industry as well as it will help to achieve value-based results by using the suggested methodologies.

CHAPTER 1

INTRODUCTION

"Organic farming appealed to me because it involved searching for and discovering nature's pathways, as opposed to the formulaic approach of chemical farming. The appeal of organic farming is boundless; this mountain has no top; this river has no end."

—Eliot Coleman, pioneer organic farmer, author of The New Organic Grower and Four-Season Farm

CONTENTS

The key to achieving a balance between the preservation of environment and maintenance of prosperity in the world is the concept of sustainable economic growth in the developed world while attaining economic development in underdeveloped countries. This chapter provides both a summary of the idea of a sustainable food industry, and a series of thorough studies of different aspects of sustainable growth and development concepts. This chapter will give an overview of the ecological effects of the food processing as well as the sustainable solutions for the food industry.

1.1. INTRODUCTION

Every individual on this planet is affected by the food supply chain. As a consequence, sustainable development of the food supply chain is important. Sustainable development has been well-defined as fulfilling "the needs of the present without compromising the ability of future generations to meet their needs" (WBCSD, 2000). The features from the production of the food, processing, supply, consumer purchase, consumer use and end of life are all included in the food supply chain, which is additionally known as the food industry or food system. The sustainability of supply chain of the food would then specify that food is produced and consumed in a manner that cares the well-being of future generations. There are several bad and unsustainable demonstrated influences of the existing food supply chain that make it unsustainable. Such effects comprise overdependence on inputs for food production such as high-intensity animal production and production of the food out of season.

For instance, the climate change is significantly contributed by the supply chain, with agricultural production responsible for 17–32% of global greenhouse gas emissions (Bellarby et al., 2008). It has been projected that the food system consumes approximately 16% of the total energy use in the United States (Hendrickson, 1996). In addition, the processing of the food amounts to approx. 25% of all water consumption globally and 50–80% of all water consumed in industrial countries (Okos).

As a consequence, the important sustainability ideas for the food supply comprise waste, energy, air, water, biodiversity, climate, food quality, food price, food quantity, employment, food safety and employee welfare (Kramer and Meeusen, 2003) (Figure 1.1).

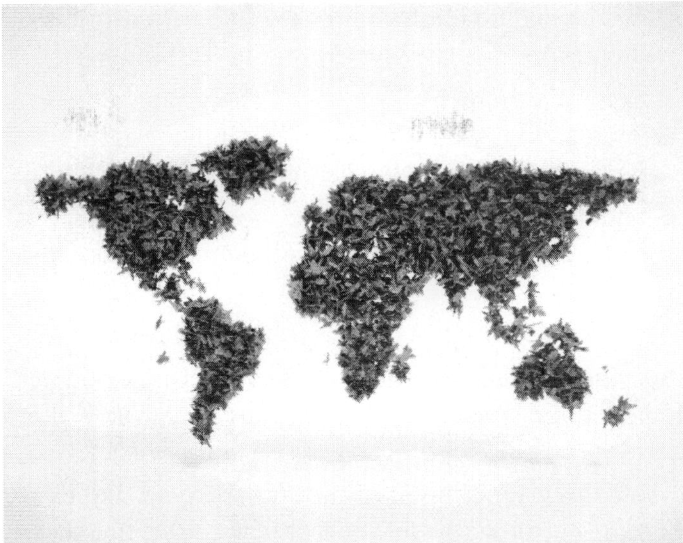

Figure 1.1: Sustainable food industry to preserve environment.

Figure: http://www.eurmscfood.nl/improving-energy-efficiency-for-food-industry-sustainability/

All these issues along with some more are discoursed thoroughly in this book. The food industry has the ability to make available harmless, nourishing, and tasty foods to a variety of consumers. Agricultural production can make available a variety of supplies for nourishment. The dispensation of supplies can offer a method to preserve foods for suitable dispersal and storage and likewise may lessen total waste by making commodities consumption. Food can now be delivered to those in need made possible due to distribution. The approach for Sustainable Farming and Food helped in development of key principles for a sustainable food chain (DEFRA, 2006), which are as follows:

1. Safe production;
2. Healthy harvests in response to demands of market;
3. Guarantee that all consumers have admission to nourishing food and to accurate data about food products;
4. Backing the feasibility and variety of rural and urban economies and communities;
5. Enabling practical livelihoods from sustainable land management, both over the market and over the payments for public benefits;

6. Consider and function within the biological restrictions of natural resources (particularly water, soil, and biodiversity);
7. Attain reliably high standards of environmental performance by dropping energy consumption, minimizing resource inputs, and using renewable energy wherever imaginable;
8. Guarantee a safe and hygienic working environment and high social welfare and training for all employees associated with the food chain;
9. Attain reliably high standards of animal health and welfare;
10. Sustain the supply obtainable for increasing food and supplying other public benefits over time, apart from where alternative land uses are vital to meet other requirements of society.

As an outcome, agricultural production should be emphasized on offering the most nutritionally solid choices with the smallest force. Food processors and manufacturers are required to comprise sustainable actions like waste reduction and recovery, recycling, composting, and dispensation with minimal water and energy use. Circulation should be as well-organized as conceivable. The welfares of sustainable practices are significant for the global social and environmental benefits stated, but the World Business Council for Sustainable Development has also discovered that businesses that integrate sustainable practices have had better financial achievements (WBCSD, 2002) (Figure 1. 2).

Figure 1.2: Environmental health.

Source: http://khortoom.com/ensha-about-environment/

There are several advantages of sustainable practices comprising lower production prices, enhanced product function and quality, bigger market share, better environmental performance, upgraded relationships with stakeholders, and lower risks. Sustainable food has generated interest in consumers. This inquisitiveness has been accredited to the want to recuperate one's private and family health and safety (Sloan, 2007). Ecological reasons have remained to be a secondary advantage, and in several methods unidentified to consumers. For example, in a study led by the Leopold Center for Sustainable Agriculture in 2007, 88% of defendants perceived local and regional food systems to be rather harmless or very safe and had acquisition preferences for such food, compared to only 12% perceiving global foods as safe (Pirog and Larson, 2007). The survey also presented that the respondents, however, did not tell that airplane transport of the food released more greenhouse gases than trucks (on a per pound basis of product transported) (Pirog and Larson, 2007) (Figure 1.3).

Figure 1.3: Unsustainable food wastage.

Source: https://www.kstatecollegian.com/2015/09/24/opinion-societys-unsustainable-food-waste-strategy-doomed-to-fail/

This book will give a brief understanding of the effects of an unsustainable food supply on health and food safety and go well past the average consumers' information. This specifies that as consumers study further, their attention in sustainable food will continue rising. Additionally, it is by now obvious that environmental concerns are moving up in importance to numerous consumers. This book will assess the sustainability of each of the chief supply chain mechanisms of the food industry. There will be importance on environmental thoughts given its implication and requirement for development. Finally, one chapter will bring the conversation from all the supply chain mechanisms together to plan sustainability principles for food and beverage products, including plans on the ways to develop more sustainable products.

The human food supply is an ongoing test of the development of a sustainable agriculture and food system that must be an essential part of our long-term economic and environmental scheduling. Suitable food and a comfortable environment are both important to the long-term existence of our species. Research and development over a period of time, have offered a notable and even unforeseen flow in production of the food across the prime agricultural regions, particularly those with potential for irrigation. Adding to these improvements have been the unexpected contributions of plant breeding to high-input production systems and the equivalent developments in fertility and pest management.

Figure 1.4: Green revolution.

Source: http://cityopen.ru/?p=11238

The benefits of the Green Revolution and the influences of the International Agricultural Research Centers provide indication of things an attentive public domain program can attain. At the same time, such an increase in the rate of the food production has come at a price. As with any organic population, human numbers have improved in answer to existing food and other resources. Human population is expected to reach the current projection of 9.6 billion before it is projected to level and drop (Brown, 2008) (Figure 1.4).

The growing human population and difficulties for food, fuel, and other products that rest on nonrenewable natural resources have put an unparalleled burden on the global life-support system (Tilman et al., 2002). Human activities presently exploit over 40% of total net primary productivity captured by photosynthesis, leaving just over half for the preservation of all other species. Possibly the brittleness of the global ecosystem is best exemplified by the present rate of destruction of plant and animal species. Lester Brown, the economist and author, once stated that "we are presently in the midst of the sixth major extinction event in the earth's history, the last of which occurred 65 million years ago, wiped out the dinosaurs, and was likely the result of an asteroid hitting the planet (Brown, 2008). Today's problem is the first such event that is almost entirely a result of human activity and our destruction of habitat." One of the instant economic and food system influences is the loss of fish and breakdown of the fishing industry, with 75% of commercial fish species being uninvolved at unsustainable rates (FAO, 2007) (Figure 1.5).

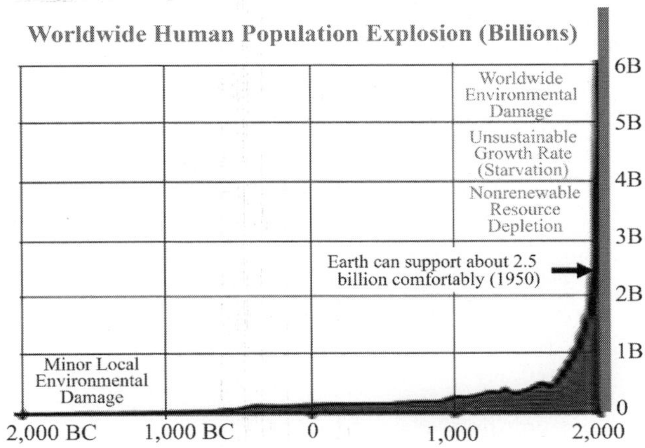

Figure 1.5: Human population growth.

Source: https://alfinnextlevel.wordpress.com/2015/07/12/can-earth-support–100-billion-humans-comfortably/

More extensive significances of these human activities comprise the losses of life-sustaining ecosystem support services (Daily, 1996). Human being must admit that their growth in human population and growth in the food production do arise at a cost, often one that they are incapable to calculate. Thus, they seem to be reaching a tilting point in the balance between manipulation of natural resources and satisfaction of human wants and requirements. Mobilizing to Save Civilization (Brown, 2008), make available a summary of present challenges together with possible solutions at the global scale.

The developments in production contributed to nutrition and better health in several extents, yet insistent poverty, particularly in sub-Saharan Africa, lasts to stops food from reaching several who most require this basic supply. The injustices in distribution of the food seem to be growing nowadays, sideways with a twisting of the economic situation between rich and poor, North, and South, all in the appearance of the food surplus in chosen areas. The existing move toward biofuels, particularly ethanol from maize and Agriculture rapeseed in the North and from sugarcane and oil palm in the south, make available another contest to food production and obtainability (Figure 1.6).

Figure 1.6: Irrigated farmlands.

Source: https://www.agpros.com/property/platteville-co–127-acre-irrigated-farmland/

The growths in the food production have been outstanding in big part to growth of irrigated farmlands and also to augmented human productivity, based on systematization that made the farming workload lighter. This process freed labor to follow other activities in the later part of the industrial revolution. Of special position is the series of genetic developments in the major food crops that has revolutionized the Green Revolution, exclusively influencing rice and wheat production in Asia and Latin America and maize production in temperate regions. These developments have been united with large surges in chemical fertilizer application and use of chemical pesticides.

The mixture of these mechanisms in well-made and effective cropping systems has produced synergisms among the innovative technologies to increase food production. Genetic upgrading of crop diversities has been a incessant procedure since the initial people settled in everlasting communities and started to extract the most necessary plants from their close environment (Plucknett and Smith, 1986). They protected seed of cereals and pulses and circulated cuttings from trees and vines that proved most needed for their home diets. Often, they were plants with the main edible seeds, those that held strongly in the head or ear or pod rather than devastating and dispersing, and those with the best cooking qualities. Most of the genetic development in improving yields of crop plants was attained by women who found these early selections, while men were off hunting, and human beings have continued to fine-tune their efforts over the past centuries.

Rediscovery of Mendel's principles of genetics was crucial to considering the procedures of crossing and growth of hybrid maize. Genetic choice methods for significant self-pollinated cereals such as wheat and rice were equally positive. International agricultural research Centre and their partners contributed in national agendas and these were essential to genetic growth in major crops all through the last half of the twentieth century. Plants make available three-quarters of the calories and protein to fuel human diets, and it is valuable to search the developments in the three major cereals, since together they contribute over half of all the energy in our food on a global basis.

The major issue which the food industry faces is the sustainability for all stakeholders who participate in the food supply chain, from agriculture, ingredient manufacturing, product manufacturing, packing, and distribution, to the consumers. The food industry has ecological responsibilities to preserve ad protect biodiversity, reduce water consumption, and reduce waste creation and emission gasses, together with caring for livelihood in terms of health and wealth local civilizations. In addition, the worldwide food industry is facing growing pressure about raw materials, ingredient tracking, and food production to provide food to the increasing population, in an environment of continuous optimization and supply chain control.

As a minimum, in order to fulfill the present global population's daily requirement necessitates the annual production of nutrients to send 360 million tons of carbohydrates, 364 million tons of protein, and 195 million tons of fat. The agriculture and fishery industries are overextended nowadays and the projected increase in the population, of 38% in the next 15 years, will mandate clever approaches (Figure 1.7).

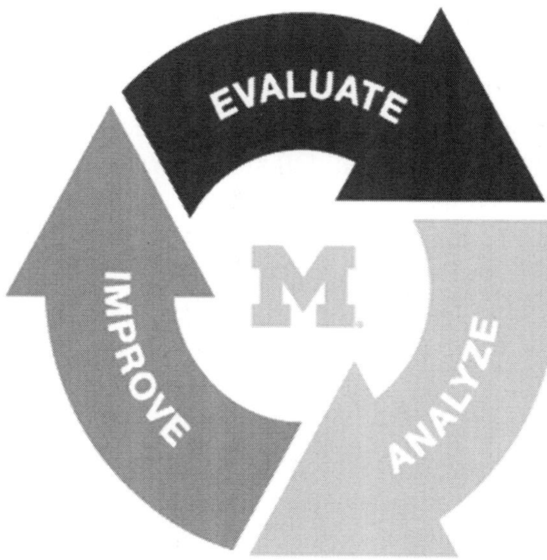

Figure 1.7: Sustainability measures.

Source: http://home.isr.umich.edu/releases/landmark-um-effort-measures-sustainability-culture-on-campus/

This book estimates industry executives' views on sustainability problems and a focus group of consumers have been interviewed about their apprehensions and sentiments. This book completes that sustainability measures have mostly been attentive on the raw material supply chain, problems associated to production energy costs, and better environmental conservation, reliant on a stakeholders' situation in the supply chain, together with individual market circumstances. The latest contests coming forward for the industry are mostly the management of biomass waste in terms of following innovations of by-product ingredients, as well as handling the general biodiversity issue on raw material usage. This book looks into existing framework circumstances for the complete food industry supply chain, the evaluation of urgencies and issues from industry leaders to consumers and estimates important measures of sustainability policies executed within the top ten global ingredient and food product manufacturers.

All the topmost multinational food ingredient and food product manufacturers have tactical objectives and operational measures ready to work for additional sustainable food production, but numerous tasks exist. Specialty food ingredients manufacturing necessitates equipment and answers to progress resource competence along the whole of the food value chain.

1.2. MARKET RESEARCH OF THE FOOD INDUSTRY

An expert research panel in June 2014 discovered that 98 industry managers regarded general performance parameters like "increase market share," "improve customer satisfaction," and "grow profit levels," as the utmost significant goals to be realized over the next three years. However, it is obviously evident that at the very top of the companies, 69% of managers managing the company's long-term existence, express that they and their administrations "Have a duty to protect the environment in which they operate." Amid all the European customers, there are regional differences concerning the rate of transparency in supply chains (Figure 1.8).

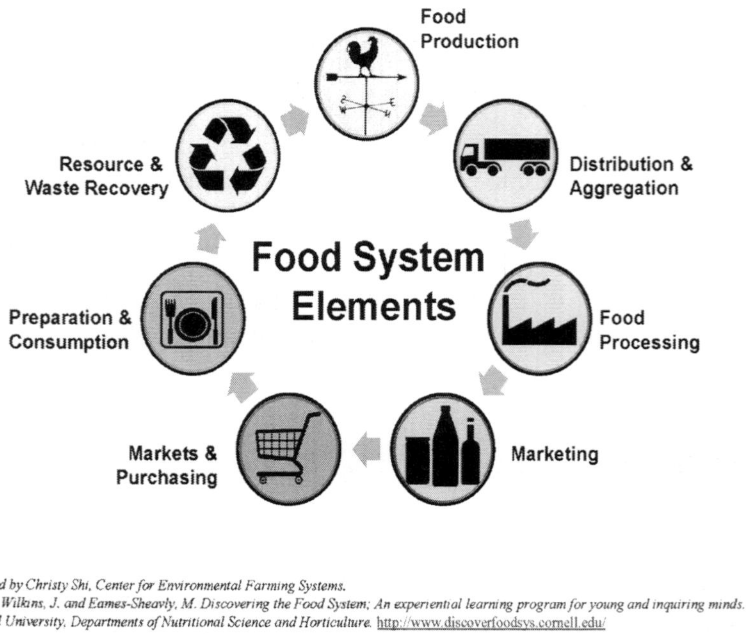

Figure 1.8: Food supply chain.

Source: http://speechfoodie.com/food-supply-chain/

Though the general consequences from UK consumers realized a steady judgment from 2013 to 2014, as 64% of all respondents measured themselves to be sustainability-aware customers, and 36% did not deliberate this to be significant. Within the international industry players, there are various important subjects put up by ingredient manufacturers, linked to processing, expenditure, water, recycling, and cleaning, and biomass management in all-purpose, via biofuel choices or the detection of innovative by-product prospects. But, ingredients by volume and life cycle assessment agendas for top agricultural raw materials are excessive on the agenda for most important ingredient suppliers, as well as R&D investments for offering extra food solutions to the increasing world population (Figure 1.9).

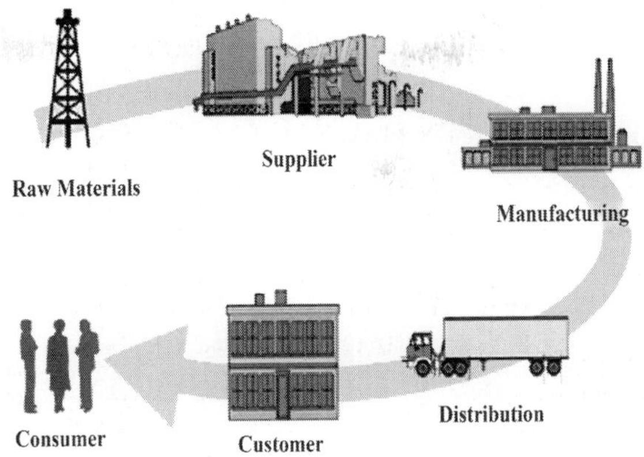

Figure 1.9: Raw material supply chain.

Source: https://futureprocurement.net/supply-chain-management/

The world's largest beverage and food producers over the years have mostly emphasized on the raw material supply side in their sustainability programs, so that they can accomplish new goals; this is in addition to the present conventional goals concerning greenhouse gas emissions and packaging reductions. Several of the international business are right into the sustenance programs of local farming communities in terms of supplying goods, accessing on problems such as seeds choices, crop rotations, fertilization, and cultivation, the group and management of crops, and even microloans prospects. The farming communities maintained characteristically grown commodity crops such as coffee, oil palms, cocoa, fruits, or vegetables. These agendas are naturally in partnerships with ingredient manufacturers, NGOs, or other interest organizations (Figure 1.10).

There are several companies which are profiled in this food industry strategies research. This comprise ADM Company, Royal DSM NV, Tate, and Lyle PLC, Kerry Group PLC, Cargill Inc., DuPont Danisco A/S, Ajinomoto Co Inc., Corbion NV (formerly CSM), Suedzucher AG, Nestle SA, Unilever NV, Ingredion Inc., PepsiCo Inc., Mondelez International Inc. (formerly Kraft Foods), Coca-Cola, Danone SA, Mars Inc., Fonterra Co-op Group Ltd, General Mills Inc., EX-ACT, Green Palm, Campbell Soup Company, Behind the Brands, Bonsucro, Nourish 9 Billion, Gates Foundation, Food Tank, Platform, Sustainable Sugarcane Initiative, Sustainable Agriculture Initiative and Rainforest Alliance.

Figure 1.10: Cocoa farm.

Source: http://www.monitor.co.ug/Magazines/Farming/Cocoa-growing-in-Uganda/689860–2417144-ypmr7m/index.html

1.3. THE SUSTAINABLE DEVELOPMENT OBJECTIVES

In terms of the food, the Sustainable Development Goals can be mentioned directly into four main goals: ending hunger, realizing food security, refining food nutrition and encouraging sustainable production. Below given are the top ten trends that are initiating the food-tech industry in the direction of an additionally sustainable foodscape.

1.3.1. Discoverability

Till now, labels were providing info about origin ingredients, disregarding the outcome of cooking procedures. Huge private and public structured-data set-ups have been made around the therapeutic value of drugs, while the part of diet, food, and lifestyle has been mainly underrated. The act is rapidly varying. The democratization of nutritional data is at this point to remain. The concluding objective is to attain the food security and information to the principal scale imaginable, both leveraging on sensors, study system and applications such as Scio or Tellspec, noticing the occurrence of particular nutrients for, e.g., gluten in Nima Sensor and making data infrastructures to develop a world public food-print.

1.3.2. Boosting Food Effect on Health

World of health is increasingly linked with food. From functional food to probiotic beverages, edible solutions that benefit in preventing or curing several pathologies are rapidly developing. It is called bio-informatics trend and the objective is to expose food causes, offering natural, sustainable, and methodically proven health solutions. In Ireland, Nuritasis making use of artificial intelligence and DNA analysis; Chloè Rutzerveld's Digestive Food project strategies food in the provision of the digestive system, using digital production. The strategy of a new food system plays a vital part in diminishing food consumption while boosting its nutrient dose.

1.3.3. Waste Reusing

When talking in relation to the food sustainability, waste is one of the top issues. In the European Union only, approximately 90M tones of the food are damaged per annum. Business prospects in this part are massive and several companies are seeming to subsidize to explaining this issue. Food Cowboy links American food producers and transporters with contributions to lessen waste; Food Cloud is undertaking the similar in Dublin, having protected 1200tones of the food till present day. Last Minute Sotto Casa (Italy) is offering sellers the occasion to provide expiring products to customers who can save money in this method. Bio Bean in UK, is reprocessing the waste from coffee grounds into progressive biofuels and biochemicals. Winnow is serving specialized kitchens to minimalize waste along the chain. The section is vivacious and the latent in sustainability impact is immense.

1.3.4. Food Production

3D food printing technologies mixes technology and digital gastronomy techniques to produce fully customized food products, unlike robotics-based food manufacturing technologies which are planned to mechanize manual procedures for mass production. This presents innumerable openings: from restructuring cooking activities such as 'what Natural Machines is doing with Foodinii.e. the first 3D food printer' to making innovative shapes beyond the bounds without these techniques such as 'what TNO is doing with pasta,' to leveraging precision nutritional openings.

1.3.5. Nerd Farming

Technology has enabled the ways for new farming prospects in indoor, warehouse-based situations. Food production can be equipped with new tools to put up high-density urban living and continue food security in spite of a future of growing climate unpredictability and susceptibilities. There are a lot of companies which are working around this area and the Open Agriculture Initiative (MIT Media Lab) is certainly one of them. Their task is to create additional farmers for the future of the food manufacturing by evolving open source hardware and software platforms for sensor-controlled hydroponic and aeroponic agriculture systems.

1.3.6. Low Ecological Effect Foods

As the world population is growing and moving towards a value of 9 billion people, the food industry is pressurized to discover innovative, ecologically friendly food answers. One of these solutions is edible insects. They comprise amino acids, high-quality protein, and vitamins for humans even though having a small influence on the environment. There are a lot of companies who are employed in this dimension. In Italy, the education about this topic is spread by Entonote; EntoCube provides technology to farm insects and insect protein ingredients for the food industry; while Bugsolutely manufactures and distributes cricket-based pasta. one example comes from algae: every bite of spirulina comprises more protein and more iron than 20% fat ground beef. The Algae Factory is transforming microalgae in attractive and sustainable food items, such as breakfast flakes and chocolate bars.

1.3.7. Food Education and Groups

"It's essential that we arm future generations with the life skills they urgently need in order to lead healthier, happier, more productive lives." quoted Jamie Oliver. The awareness of the food and its education is a trend that is on a rise steeply and additional administrations are attaining the rank of creating "food hubs and online communities" from one place to another in the world, which are ultimately linked. Some of them which are worth mentioning are Food + Tech Connect in NYC, The Food Tank in DC, the Jamie's Food Revolution community and Future Food in Italy. Knowledge is demonstrating to be one of the record effective weapons to form a sustainable system.

1.3.8. Small & Reasonable Food Supply Chain

A very few intermediaries ate involved in short food supply chains, they improve the rural economy, generate innovative methods of vending local products and fascinate new types of clients. This appears important in the aspect of the mounting global population. Businesses who are attempting to style it as simple as likely for the ultimate customer are dropping in, undertaking the subject from different viewpoints. Slow Food and its youth system is a wonderful instance. A global grassroots movement of the food producers, farmers, scholars, and food professionals that are vigorously endorsing good, clean, and fair local food systems around the world (Figure 1.11).

Figure 1.11: Fair and good local food system.

Source: https://www.tes.com/lessons/kFWbUdpogkRAbg/food-systems

1.3.9. The Easiest Choice Must be the Healthiest Choice

A significant objective while conversing about healthy and sustainable food is to produce the healthier food selection the easiest one. Michiel Bakker of Google Food clarified in what way his business is attempting to do that every day for its personnel, employed on food as a product in addition to the technique it is located and helped to Googlers. In the meantime, numerous startups are displaying up with exciting healthy product concepts which could possibly substitute existing food with better and more sustainable solutions. Some examples from frecious, offering high-quality and ready-to-eat, nutritious, and indulgent, sustainably made and conveniently packed spreads.

1.3.10. Smart Cookbooks

As soon as people get to know about the things to eat, they need to be aware of the ways to eat it. In these how-to-add-it instances, they're searching for diverse forms and recipes. The most critical thing while talking about nutritional sustainability is In. A motivating instance in the space is Food pairing. "Working with aromas (80% of what we call taste is actually aroma) they are able to profile specific ingredients through gas chromatography coupled mass spectrometry (GC-MS)," and couple them over data analysis and machine learning. The objective is to allow everybody to produce the best food picks in terms of taste and nutritional value.

1.4. SUSTAINABILITY SOLUTIONS FOR THE FOOD INDUSTRY

The complication of the food production, global food supply chains and distribution, promising the sustainability and environmental effect of the food and drink products can demonstrate a contest. Intertek's sustainability experts back producers, suppliers, and retailers to fulfill their sustainability and controlling goals.

Intertek benefit provision food and drink client administrations to grow custom-made programs to accomplish their supply chain, energy, disposal of waste and water footprint attaining a greener, more effective manufacturing and distribution procedure.

1.4.1. Life Cycle Analysis for Food

The life cycle analysis for food helps them to grow custom-made programs to be able to handle their supply chain, energy, disposal of waste and water footprint achieving a greener, more well-organized manufacturing and distribution procedure. For the food industry, the most complete method for evaluating the complete environmental effect of the food and beverage products is Life Cycle Assessment (LCA) from raw ingredients through to processing and manufacture, delivery, use, waste management and final dumping. Intertek sustainability experts can assist in understanding the ecological effect of any product at each stage of its life cycle and can help businesses in reducing operational costs by prompting, arranging, and reducing their ecological footprint at the same time as fulfilling consumer demand for developments in sustainability (Figure 1.12).

Figure 1.12: Life cycle analysis for food.

Source: http://kpadltd.co.uk/area-of-expertise/food-life-cycle-assessment-lca/

1.4.2. Carbon Footprint of the Food Manufacturing and Products

Main sustainability actions for the food sector can be evaluated by making use of industry-accepted practices and procedures, together with the Global Reporting Initiative (GRI), the world's most widely used sustainability reporting framework, and the Greenhouse Gas Protocol, an accounting tool for understanding, measuring, and handling greenhouse gas emissions (Figure 1.13).

Figure 1.13: Carbon emission.

Source: https://en.wikipedia.org/wiki/Greenhouse_gas_footprint#/media/File:Apollo_Beach_power_plant_01432.jpg

Intertek's Sustainability experts can help you to comprehend the influence that any food and drink business has on the weather and work with them to recognize the most active method of falling costs and emissions, whether it is in owned operations, supplier operations or in the customers use and disposal phase of products which may help them in building long-term and carefully competitive relations with suppliers.

1.4.3. Regulation of the Food Product Supply Chain

Intertek offers a single-source explanation for companies and their suppliers to fulfill and obey the global green regulatory necessities through all-inclusive valuation, product testing, authentication, product guarantee, and consulting services to support companies positively attain green supply chain management from achievement of raw materials, to production and delivery of finished goods to customers or retail outlets.

1.5. SUSTAINABLE FOOD MANUFACTURING DRIVERS

The main drivers in the direction of achieving sustainable food processing are economic, consumer, legislative, or corporate based (Figure 1.14).

Figure 1.14: Main drivers of sustainability.

1.5.1. Legislative Drivers

According to Cooke (Cooke, 2008), environmental legislation related to the food manufacturing industry can be measured at three levels (Figure 1.15):

- at a point in the situation;
- at the point of emission; and
- at the procedure that generates the emission.

Figure 1.15: Pollutant emission.

Source: https://www.eea.europa.eu/highlights/air-pollutant-emissions-declining-but

In wide-ranging, there has been a direction to the third level, that is, controlling the process that releases the pollutant, before emissions proved harmful to the environment. In Europe, such control attempt appeared as the Integrated Pollution Prevention and Control Directive (Europe sources all over the European Union, as well as the food and beverage sector. This is accomplished through the establishment of a "license to operate" to sites that fulfill certain circumstances. Licenses must consider the entire environmental presentation of the plant, together with aquatic, air, and solid waste generation, use of raw materials, noise, prevention of accidents, energy efficiency, and refurbishment of the site on conclusion (European Commission, 2010).

Additionally, notable is that emission limit values in the permit must be based on the use of the "best available techniques" (EU Joint Research Centre, 2013). They must perform the full examination of environment facility and the public has the right to be involved in the procedure of decision-making. Hence, it can be guaranteed that food-processing services, through design or promotion, encounter international environmental standards.

Consequently, at a very early stage, innovative facilities must be planned to fulfill these environmental necessities, while older plants may require to be retrofitted with modernized technologies. In the European Union, member states are lawfully enclosed to enhance the treatment of bio-waste. The "waste hierarchy" discloses that the prevention of waste is the best choice, followed by recycling, reuse, and energy retrieval (Figure 1.16).

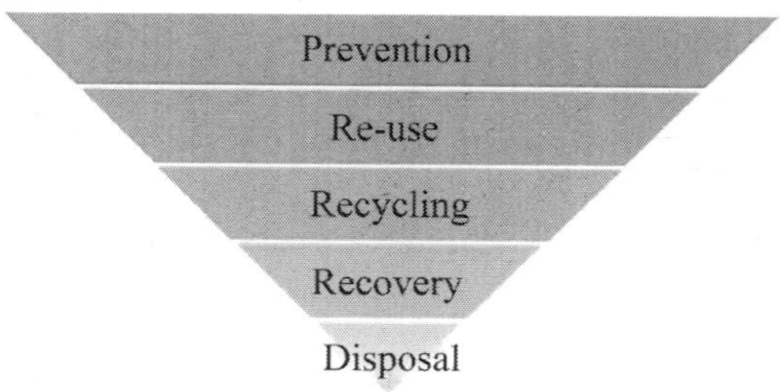

Figure 1.16: Waste hierarchy.

Source: https://www.recycle-more.co.uk/business-zone/office-recycling-and-waste-management/the-waste-hierarchy

The worst option for environment is disposal method, such as landfilling. This has led to legislation necessitating member states to decrease the amount of bio-waste they direct to landfill (European Commission, 1999). Thus, options for waste disposal by food processors are inadequate, mainly in countries such as Ireland, which were formerly seriously dependent on landfill. This has played a part in the increasing interest in substitute use and disposal choices for waste streams in the food processing sector. Further, this will be discussed in the next section. It must also be taken into account that there is a transfer in the direction of connecting legislative drivers with economic drivers to progress overall sustainability.

For example, the compulsion for food and beverage companies is EU Emission Trading System (ETS) operating incineration fittings above a capacity of 20 MW. Instances of amenities which have dropped under phase III of the ETS comprise those comprised in snack food production, distilling, and brewing, butter, and cheese production, and coffee processing (UK Department of Energy and Climate Change, 2012). The ETS is a "cap and trade" scheme, which has objectives of launching an allowance-trading arrangement for emissions, and indorsing decreases of greenhouse gases, in specific carbon dioxide. It is mandatory for every EU country to give in to National Implementation Measures (NIMs) for sanction by the European Commission. NIMs regulate the levels of free distribution of allowances for installations throughout the present phase of ETS. These distributions are built on EU community-wide synchronized rules (European Commission, 2011). Over the progress of a carbon market, companies can purchase payments to fulfill their compliance necessities or sell their surplus allowances. Businesses purchase distributions through auctions. Two auction platforms are by now in place. The common platform for the big majority of countries taking part in EU ETS is European Energy Exchange (EEX) in Leipzig. The EEX also acts as Germany's auction platform. The second auction platform is ICE Futures Europe (ICE) in London, which acts as the United Kingdom's platform (European Commission, 2013).

1.5.2. Economic Drivers

A number of forms can be taken by economic drivers in sustainable food processing. It has been anticipated that world promoted energy consumption will produce by 53% between 2008 and 2035 (EIA, 2011). This, joined with an anticipation that energy prices will last to growth in the long term (EIA, 2011), has turn out to be a driver for companies, comprising food processors, to decrease energy consumption (Figure 1.17).

Figure 1.17: Sustainable food processing.

Source: https://beingreenz.wordpress.com/2013/06/07/sustainability-of-food-systems/

Carefully chosen unit operations in the food processing amenities are chiefly energy exhaustive, for example, drying, and evaporation. It has been recommended that energy can make for up to 10% of the total production costs for products needful these unit operations (Marechal & Muller, 2008). Although this number is low in evaluation with other industries (oil refining ~60% of total production cost), increasing energy costs have required energy saving and energy management programs to the forefront of many food processing businesses. Waste material of range between 30% and 50% from incoming raw materials can be generated during food processing (Henningsson et al., 2004; Schaub & Leonard, 1996). In the UK, food-producing units produces roughly 2.5 billion kg of the food waste annually (Hall & Howe, 2012); in the US, the EPA evaluates that it costs around $1 billion to dispose of the food waste (Kosseva & Steve, 2009).

In the EU, a major worry is a landfilling of bio-waste, with around 40% of waste end up in landfill. As mentioned in earlier section 1.4.1, there is a legislative obligation for EU member states to decrease the use of landfill as a waste disposal selection. To inspire this, governments have improved the duty on landfill to make available an economic incentive for companies to transfer to additional sustainable approaches of waste disposal. Factually, some food waste watercourses would have been used as animal feed supplements, e.g., whey as pig feed. Though, with the Food Processing: values and claims acknowledgement of the added value of whey components, nowadays the trend is to make use of it in an additional sustainable manner (Figure 1.18).

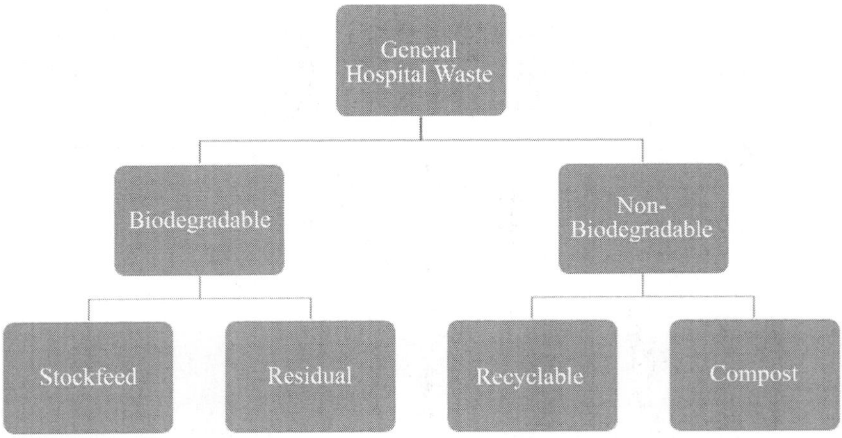

Figure 1.18: Bio-waste streams.

Source: https://www.researchgate.net/figure/Hospital-Waste-management-Flow-Chart-Classification-of-Bio-Medical-Waste-The-World-Health_fig1_270745760

Therefore, as an outcome of the force to decrease waste and utilize it more sustainably, there has been an improved emphasis on making use of the food processing bio-waste streams to grow waste-to-energy systems and value-added products, in addition to discovering new, more sustainable, waste disposal choices.

1.5.3. Consumer Drivers

The consumer power and demand from the consumer can drive developments in the market economy where it could lead to sustainability in the food processing industry. Various studies have found out that consumers will distinguish products based on their sustainability (Jaffry et al., 2004; Kearney, 2010). Though, consumer choice is multifaceted and will be affected by a group of factors together with cost, convenience, and habit (Vermeir & Verbeke, 2006). It is assessed that about 10% of the population are "socially conscious purchasers" (Sahota et al., 2009). Product entitlements can variety from the utilization of recyclable material and product lifecycle sustainability, inclusive of company sustainability obligations. The part of certified products, i.e., products confirmed by a third party to fulfill a set of standards, in endorsing sustainable food processing through consumer-driven demand is increasing (Figure 1.19).

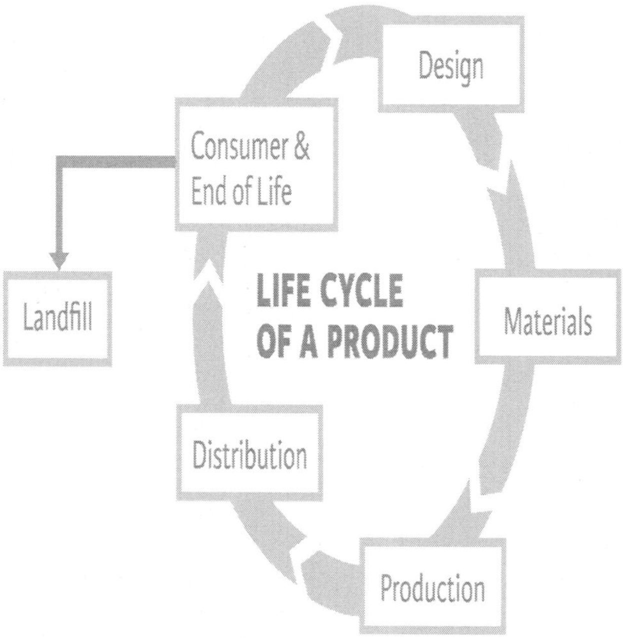

Figure 1.19: Product life cycle.

Source: http://www.ethicsingraphicdesign.org/morality/sustainability/

Certification can have a supervisory part, such as certification essential by the EU or US Food and Drug Administration (FDA), or can happen over voluntary schemes, e.g., Rainforest Alliance or Fairtrade. Arrangements such as Fairtrade and Rainforest Alliance emphasis on the establishment of sustainable livelihoods and/or preserving ecosystems and biodiversity in spaces where products or raw materials initiate, while other schemes, e.g., Mobius Loop, specify that a product or part thereof can be recycled where facilities are obtainable. For voluntary certification schemes to offer extreme advantage to food processors, rights are essential to be trustworthy, easily understood, and trusted by consumers. A number of research have examined consumer purchasing selections of "green labeled" products. Jaffry et al.

(2004) examined consumer choices for quality and sustainability-labeled seafood products in the UK (Figure 1.20).

Figure 1.20: Sustainability labeled seafood.

Source: https://www.hakaimagazine.com/news/flawed-promise-certified-sustainable-seafood/

They interviewed 600 in-home people and discovered that certification signifying sustainability or quality had a meaningfully optimistic effect on product choice, with sustainability seeming to have the highest encouraging impact on the probability of choice. However, they similarly discovered undesirable attitudes detained in the direction of non-governmental certifiers, over governmental certifiers, stressing upon the status of consumer trust in certification schemes. While such research shows the possible worth to food processors of implementing green labels, it is also serious to confirm that consumers identify, comprehend, and value on-package info on production standards (Hoogland et al., 2007). Research has revealed that when food products (chicken, salmon, milk,) are considered with either

1. A certified organic logo united with facts about the animal welfare standards (e.g., easy stunning/ murder, outside admission, organic fodder, etc.),
2. Just the certified organic logo, or
3. A declaration in which the produce was credited to the world market (e.g., follows to legal production standards), several customers were unsuccessful to understand that the organic logo

offered alone (option b) enclosed all three categories (Hoogland et al., 2007).

In fact, they were persuaded to undervalue the characteristic benefit of the logo and were more optimistic towards products that offered both the logo and particulars of animal welfare.

1.5.4. Corporate Performance

In general, food-processing companies are trying to solve environmental subjects. Several businesses are now counting corporate performance as a measure of sustainability (Pirog et al., 2009). There are three "dimensions" of sustainability:

- Environmental sustainability, integrating the managing of the impacts of human activity so that they do not enduringly damage the natural environment;
- Economic sustainability, which includes handling the financial dealings linked with human activities so that they can be continued over the long term without sustaining intolerable human hardship; and
- Social/cultural sustainability, i.e., permitting human actions to progress in such a method that social relations among people and the numerous dissimilar cultures around the world are not unfavorably pretentious or permanently tarnished (Figure 1.21).

Figure 1.21: Dimensions of sustainability.

To be a successful truly sustainable business, it must prosper in all three categories. This has to turn out to be known as the "triple bottom line." Environmental sustainable corporate performance (SCP) is well-defined as

"Good housekeeping through prevention of pollution and waste and efficient use of scarce resources" by Gerbens-Leenes. The contest for companies can frequently be by what means precisely evaluate their performance in each of these three criteria. Often sustainability and ecological problems in the food processing companies emphasis on local-level eco-friendly effects linked with their operation by means of a large number of indicators (Gerbens-Leenes et al., 2003). Hence, valuation of sustainability below all three pillars at not only a local but a regional and global scale may be misplaced.

Generally, the increase in demand for high-quality food, attached with the circumstance that the ecologically aware consumer of the future will reflect ecological and ethical criteria in choosing food products, will, in combination with legislative, economic, and corporate drivers, play a role in the growth of really sustainable food processing (Roy et al., 2009).

1.6. ECOLOGICAL EFFECT OF THE FOOD PROCESSING

To convert the raw materials such as fruit, vegetables, milk, and cereals, etc. into food which is fit for consumption by human being, the food processing is used. In additions to raw materials, the real contributions to the system are water and energy. Hence, the principle environmental effects of the food processing sectors are oceanic, climatic, and solid waste generation, which are impacted by the amount of resources used, waste created, and transport utilized as a part of the processing system. Fruitful administration of water and energy resources in the food processing is critical to cost-effective sustainability rehearses. Environmental sustainability can be accomplished by creating and executing alternate environmental best-practices innovations and items that boost the productive utilization of resources and accomplish cost savings, while limiting negative human and environmental effects (Clark, 2010) (Figure 1.22).

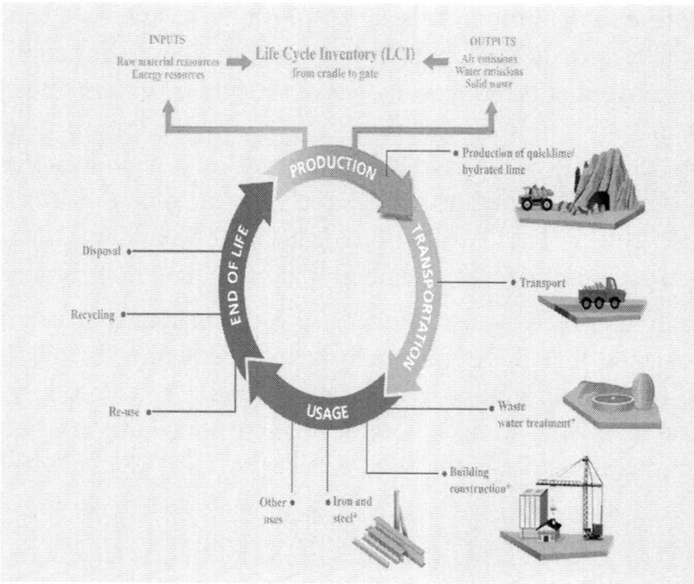

Figure 1.22: Life-cycle-assessment.

Source: https://www.eula.eu/topics/environmental-impact-assessment-lca-life-cycle-assessment

1.6.1. Energy

The food industry is a noteworthy source of air emissions, chiefly caused by broad energy use. The food industry consumes a substantial amount of energy for heating building, in processing and giving processed water, and for refrigeration and the transportation of raw materials and items (Pap et al., 2004). Such energy use adds to ozone-depleting substance (GHG) emissions, particularly carbon dioxide emissions (Roy et al., 2009). In spite of the fact that GHG emissions are a specific challenge for the food production sector, which depends intensely on transport and manure use, energy depletion and administration in the food processing have additionally turned out to be progressively imperative, as talked about in segment given below, because of economic and legislative pressure. Over the recent years, the food industry has gained ground in decreasing its energy utilization through process optimization and control, energy restoration and reusing systems, and good manufacturing practices. This pattern is probably going to proceed because of the requirement of enactment in carbon exchanging frameworks, e.g., EU ETS (Sellahewa and Martindale, 2010) (Figure 1.23).

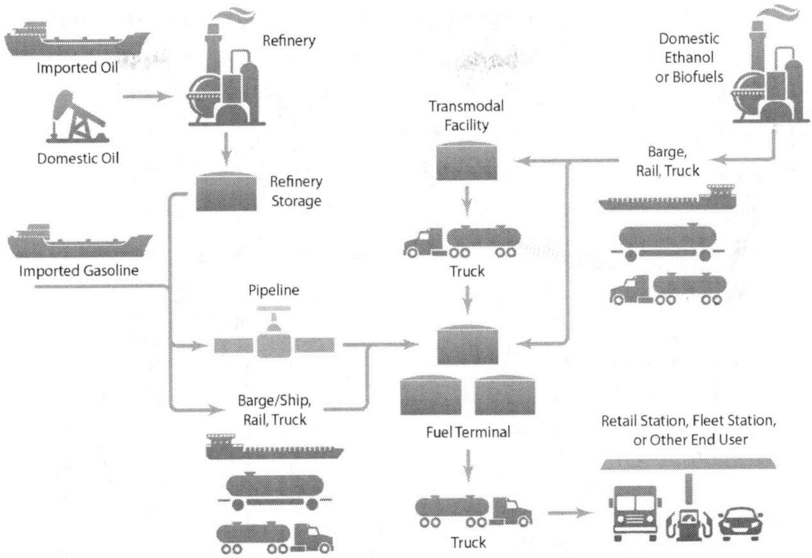

Figure 1.23: Energy productive system.

Source: https://www.afdc.energy.gov/fuels/ethanol_production.html

Advancement of energy-productive system design is, today, standard in the food processing industry. Recovery is utilized in procedures, for example, milk pasteurization to accomplish considerable energy savings. Raw milk entering a plate warm trade is preheated to ~55 C by hot purified milk on the opposite side of the plates; the raw milk at that point moves from the recovery area to the following segment of the plate warm exchanger, where high temp water is utilized to warm the milk to the required processing temperature of 72°C. The utilization of recovery lessens the measure of high temp water required to warm the milk to the purification temperature, along these lines diminishing the steam/energy prerequisite of the procedure. By using recovery, purification is 95% effective at recuperating heat. Multi-effect designs are likewise standard for procedures, for example, evaporation.

Multi-impact evaporation is evaporation in different stages, whereby the vapors produced in one phase fill in as a heating source to the following stage. In a multi-effect evaporator, for every kg of water evaporated, the amount of steam consumed is inversely proportional to the quantity of impacts and the amount of cooling water used in the condenser is inversely relative to the quantity of impacts (Berk, 2009). Procedures, for example, pasteurization are clearly highly effective in term of heat recovery. In any case, in the UK it is still assessed that around 2.9 TW/h of recoverable heat is wasted yearly by the food and drink processing area (Law et al., 2013) (Figure 1.24).

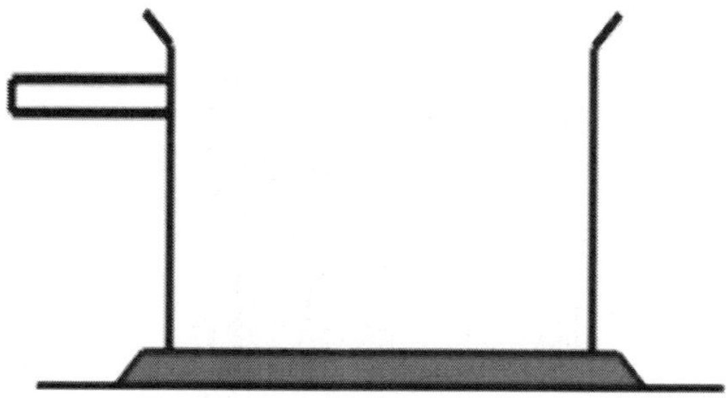

Figure 1.24: Evaporation.

Source: https://commons.wikimedia.org/wiki/File: Evaporation.png

The effective capture and use of such waste heat can possibly assume a basic part in enhancing the general energy productivity of this division. It is evaluated that the recovery of this waste heat could possibly bring about cost reserve funds of £70 m and emissions investment funds of 514,080 tCO2eq for the UK food and drink processing industry yearly (Law et al., 2013). Meeting such targets will require various difficulties to be overcome. It is frequently hard to precisely coordinate the waste heat with suitable heat sinks, in this way diminishing the percent of heat recovered. It is likewise exceptionally hard to recover waste heat in the scope of surrounding to 60 C. Other than the pasteurization and evaporation processes specified above, different sources of waste heat incorporate broilers, ovens, dryers, coolers, boilers, power plants, and air compressors.

Uses for waste heat can be categorized as follows:

- Use within its originating process (e.g., regeneration);
- Use within another process;
- Use of cooling via absorption refrigeration cycle;
- Use for space and water heating (Reay, 2008).

Irrespective of its finish use, the waste heat source should fulfill all the food processing hygiene necessities for the procedure in which it will be utilized. Instances of the use of waste heat in other procedure comprise:

- "Recovery of waste heat from two 50 kW compressors, which is used to preheat boiler feed water, resulting in a 3% reduction in boiler energy demand at a biscuit factory" quoted by Reay.
- "Recovery of waste heat from the fryer section and exhaust stream of a potato crisps manufacturing plant is used to drive an organic Rankine cycle system for power generation and could have the potential to meet the average power requirement as well as 98.58% of the peak power requirement of the manufacturing process" quoted by Aneke.

A high electricity is demanded in conventional refrigeration. Though, the usage of waste heat to calculation of substitute refrigeration cycles can decrease this. In the absorption refrigeration cycle, waste heat can be utilized to drive a generator where ammonia is evaporated off. The subsequent vapor contains some water, so a rectifier and dephlegmator are incorporated to diminish this. This vapor stream condenses into liquid in the condenser, dismissing the heat of condensation.

Condensed liquid goes through a valve and low-pressure liquid enters an evaporator to evaporate the heat required for evaporation. The subsequent vapor at that point enters the absorption cycle and is eventually returned to the generator. While this cycle has the advantage of utilizing waste heat, its coefficient of performance is lower than an ordinary mechanical vapor compression cycle. In this way the absorption cycle can be combined with other refrigeration cycles, for example, the ejector refrigeration cycle, to enhance the general coefficient of execution.

To enhance the sustainability of the food industry altogether, technological progressions in energy efficiency should be combined with expanded utilization of sustainable power source, for example, solar, wind, and bioenergy. Organizations, for instance, Dole (2008) and Kettle Foods (2009), have begun to fuse sustainable power sources, for example, biodiesel, wind, and sun-based energy, to enhance their sustainability. Such decisions should be joined with expanded effectiveness, as featured by a recent study of the Australian prune-drying industry, which showed that up to 60% of energy could be saved by improvement and control of the procedure and using solar energy (Sellahewa and Martindale, 2010). It is likewise conceivable to convert over food waste stream to energy to decrease energy necessities and cost; this will be talked about in the below section 1.5.2.

1.6.2. Solid Waste

The food industry is a noteworthy generator of both solid waste (food, packaging, and so forth.) and fluid effluents (see segment 1.5.3) all through the processing chain. Generally, food wastes go to landfill, yet this isn't currently regarded appropriate as it represents a serious environmental concern (see area 1.5.3). At the point when food is wasted, it additionally adds to GHG emissions and water utilization, as energy and water are utilized as a part of developing the raw materials, processing the product and in storage and distribution. In a current life cycle evaluation completed with new Australian mangos, it was demonstrated that waste added to 53% of the general GHG emissions amid generation, transportation, and utilization stages (Ridoutt et al., 2010a) (Figure 1.25).

Figure 1.25: Solid waste in a landfill site.

Source: http://efc.web.unc.edu/2013/11/05/solid-waste-finance/

In this manner, food waste must be handled on different levels, i.e., processor, wholesaler/retailer, and buyer. The significant supporters of solid waste from food processing are natural product processing, cocoa/chocolate/dessert shop, fermenting/refining, and meat processing (My Dieu, 2009). For instance, poultry processing produces 35 kg of bio-waste per 1000 birds processed, while refineries deliver 300 kg of bio-waste per ton of definite item (My Dieu, 2009). Obviously, the structure of the bio-waste stream is reliant on the procedure that creates it, and its piece will, thus, decide how the waste must be prepared (Figure 1.26).

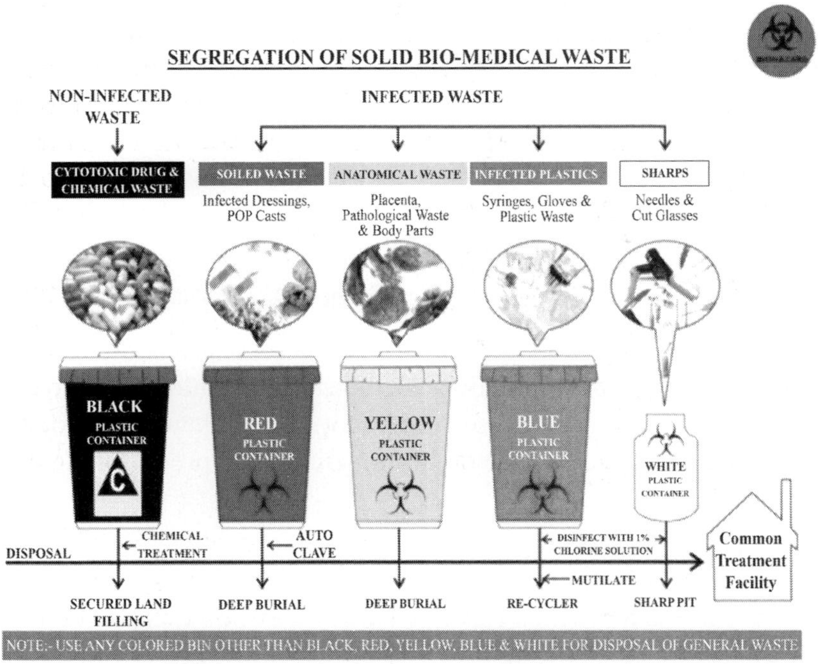

Figure 1.26: Waste management system.

Source: http://karenvis.nic.in/Database/Waste_Management_7899.aspx

A successful waste administration system will plan to diminish the volume of waste produced and reduce value components within the waste, followed by treatment and discharge. The minimization of waste generation can be accomplished through good manufacturing practices, satisfactory process control, and proper equipment maintenance and plan. In the event that any of these criteria bomb, there can be a lessening in process proficiency and an expansion in item misfortunes amid generation. Methodologies that expand time span of usability are additionally essential in lessening general food waste. Procedures, for example, food irradiation can broaden time span of usability of items, for example, potatoes (restrains growing) and strawberries (hinders development of waste microorganisms) (Figure 1.27).

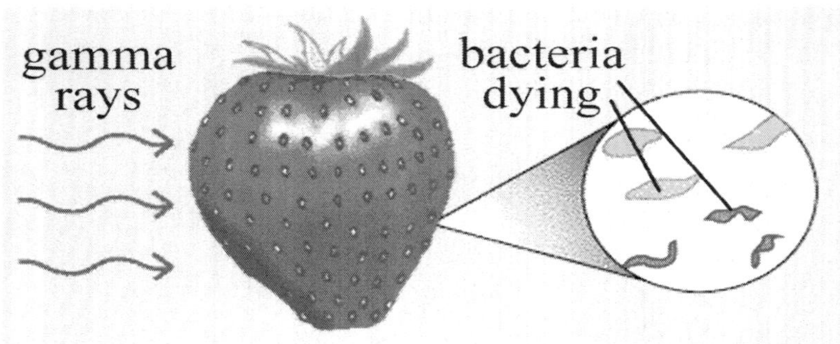

Figure 1.27: Food irradiation.

Source: http://www.teachnlearnchem.com/Food%20Irradiation.htm

Be that as it may, irradiation has not gotten across the board use. Consumer acceptance of the food irradiation fluctuates from nation to nation, which has additionally obstructed its selection. It ought to likewise be noticed that significant efforts have been made by numerous food processors to decrease food bundling.

For instance, in the UK, various retailers and processors have joined to the Waste and Resources Action Program (WRAP) and the Courtauld Commitment 2 (www.wrap.org.uk), whereby they point, in addition to other things, to decrease packaging by 10%. For quite a while, waste flows were simply seen as results of the food processing, yet today they are regularly considered as important additional revenue flows. In this way, waste flows are frequently additionally prepared so as to remove every single significant segment prior discharge. These parts are overhauled into esteem-included items. A decent example of this is whey from cheddar making.

1.6.3. Water and Waste Water

Guaranteeing the sustainability of water systems is as of now a noteworthy concern. The food industry is a huge consumer of water over the food production chain, although according to Drastig et al. (2010), agricultural production has represented around 90% of worldwide fresh water utilization amid the past century. It is likewise expected that water consumption for food production should increase to meet demands of 50% worldwide population.

Milman and Short (2008) expressed that 20% of the total population right now needs access to clean water and existing levels of access may deteriorate because of intensifying urban water infrastructure, environmental change, and environmental degradation (Figure 1.28).

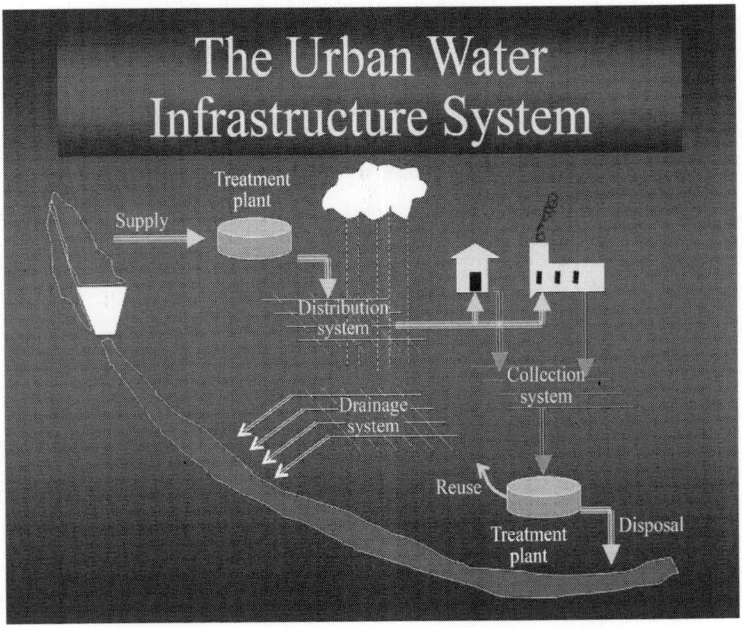

Figure 1.28: Urban water infrastructure.

Source: http://slideplayer.com/slide/6152756/

1.7. GREEN STRATEGY AND ITS NEED

The food and beverage industry have a big carbon footprint. A green strategy will positively impact the environment and your bottom line.

Due to global climate concerns, as well as a number of other environmental and consumer-preference factors, businesses have begun focusing their advertising campaigns and marketing materials around sustainability and eco-consciousness.

But adopting green initiatives starts from the ground up, and there is much more to establishing eco-consciousness as a core value than simply marketing that intention to the consumer. As the demand for food and beverage commodities continues to rise, the F&B industry must begin to adopt sustainability and environmental consciousness throughout production lines in order to mitigate the effects of higher demands.

Many of these changes require nothing more than simple innovation, but others require new strategies based on data and analysis gathered via supply chain management technologies. These technologies can help executives gain a more comprehensive view of their entire supply chain and make concerted efforts to go green wherever possible.

1.7.1. Comprehensive Analysis of Supply Chain

Every action plan should start out with a comprehensive overview from which to work. Without clear insight, proper supply chain analytics, and real-time data, inefficiencies throughout the production workflow are much more difficult to identify and address.

A comprehensive green strategy begins with a comprehensive map of the entire supply chain. From there, production can be evaluated from start to finish. Take a look at your supply chain from the ground up, looking at each of the following areas:

- suppliers;
- raw ingredients;
- packaging;
- use of nonrenewable resources;
- energy consumption;
- the use of hazardous materials;
- carbon emissions;
- landfill deposits; and
- water usage.

Once you have an overview of your supply chain, a risk assessment can be performed to identify where the focus should be placed in order to increase sustainability and enact green initiatives. Risk assessments should focus on all areas where sustainability and environmentalism may be in jeopardy, and then work from that assessment towards effective solutions.

Once areas of risk have been identified, it is much easier to go to the task of minimizing or rebuilding any negative or inefficient processes such as those that are fossil fuel-based, in order to replace them with greener solutions. Much of this information can come from a good enterprise resource planning software system that helps you manage your supply chain, interpret data, and receive clarity about the entirety of your business from product planning to sales and marketing. By analyzing data and setting up measurable goals, you can effectively demonstrate real, quantifiable change throughout your operation.

1.7.2. Increasing Efficiency

Increasing efficiency throughout production not only speeds up a product's time to market, but it also has a great impact on the eco-friendliness of a food and beverage supply chain. Reducing overhead and scaling back on facilities also lends itself to increasing the sustainability of the business. Efficiencies throughout the operation can help increase water conservation, lower fossil fuel usage, and reduce harmful emissions.

Oftentimes, process optimization just takes a little bit of innovation. Effective enterprise resource planning software can help a business review current workflows and search for sustainable solutions. From streamlining logistics to auditing usage and production, resource management technology can offer the transparency necessary to identify gaps and inefficiencies throughout. From there, the effective implementation of best practices is a much cleaner and clearer process.

1.7.3. Better Modes of Transportation

Fuel economy and emissions can be quite impactful where large fleets of automobiles are concerned. Lower CO2 emissions and better fuel economy can be obtained by simply moving to a hybrid model of automobile. Although more costly upfront, the environmental impact and fuel savings throughout the life of these vehicles can more than makeup for the upfront costs of revising your own fleet or selecting form distributors that employ hybrid vehicles.

1.7.4. Increased Diversity

By making your green initiatives known and diversifying your suppliers, adherence, and compliance can find time to shine. Competition between suppliers can help encourage them to seek out greener processes and keep their businesses legitimately in compliance with the food and beverage industry's preference for sustainable processes.

By holding manufacturers, vendors, and suppliers accountable for implementing standards, best practices can be upheld and food and beverage manufacturers can be confident that their promise to maintain green initiatives as a core value remains intact.

1.7.5. Water Conservation

Water conservation is one of the most important sustainability issues faced by the food and beverage supply industry. Often, a large amount of water is used throughout production for heating and cooling, bottling, irrigation, washing, and more.

The large volume of wastewater produced is a perfect target for efforts to increase sustainability. By using filtration and disinfection techniques to reuse wastewater produced through manufacturing processes, the environmental impact of such practices can be greatly reduced.

1.7.6. Transparency

Green initiatives are about more than public relations. The act of disclosure not only helps consumers understand and place faith in your business, but it actually helps you maintain compliance by keeping these issues clearly in the public's line of sight. This ensures that all aspects of the business are upheld and no corners are cut. With declarations of compliance with industry standard and regulations, complete cooperation with local and government bodies, and open and honest communication with the public, food, and beverage suppliers have no choice but to maintain integrity throughout the entire business from raw material suppliers to packagers and distributors.

As food and beverage suppliers begin disclosing more and more information about their labor forces, their material sources, and their practices, the quality of those practices naturally increases. This helps build reputation and security in the buyer's eye that they are purchasing products from a conscientious business.

Because of the high demand and a growing population, the food and beverage supply chain's impact on the environment is increasing. And as climate change brings about palpable change throughout the globe, the responsibility and sustainability of each food and beverage supply chain carries greater importance. Because food and beverage supply chains often have agricultural components related to use of water, employment of labor, and impacts on soil, the management of efforts to go green are complex. It requires extensive review, research, and planning in order for business owners and executives to remain true to their efforts to reach consumers through reliable, transparent, and sustainable practices.

1.8. CONCLUSION

As examined the degree to which farmers have enhanced food creation as of late with minimal effort, locally accessible and environmentally touchy practices and innovations. The study above is broke down amid 1999– 2000, projects in 52 creating nations, in which 8.98 million ranchers have received these practices and advances on 28.92 million hectares, speaking to 3.0% of the 960 million hectares of arable and perpetual harvests in Africa, Asia, and Latin America. The enhancements are discovered in the food generation occurring through at least one of four instruments:

1. intensification of a single segment of the farm system;
2. addition of another productive component to a system;
3. better utilization of water and land, so expanding cropping intensity;
4. improvements in per hectare yields of staples through the introduction of new regenerative elements into farm systems and new locally suitable harvest varieties and creature breeds.

REFERENCES

1. Baldwin, C., (2010). *Principles of Sustainability For Food Products*. [online] Food manufacturing. Available at: https://www.foodmanufacturing.com/article/2010/08/principles-sustainability-food-products [Accessed 17 Mar. 2018].
2. Future Food, (2016). *The Future of the food: Ten Trends in the food Sustainability – Future Food*. [online] Available at: http://futurefood.network/the-future-of-food-ten-trends-in-food-sustainability/ [Accessed 17 Mar. 2018].
3. Intertek.com, (2018). *Sustainability Solutions for the Food Industry*. [online] Available at: http://www.intertek.com/sustainability/food-drink-industry/ [Accessed 17 Mar. 2018].
4. Murphy, F., McDonnell, K., & Fagan, C., (2018). *Sustainability and Environmental Issues in the Food Processing*. [online] Nfscfaculty.tamu.edu. Available at: http://nfscfaculty.tamu.edu/talcott/courses/FSTC311/Textbook/9-Chapter%209%20Sustainability%20and%20Environment.pdf [Accessed 17 Mar. 2018].
5. Pretty, J., Morison, J., & Hine, R. (2001) *Reducing Food Poverty by Increasing Agricultural Sustainability in Developing Countries*. www.sciencedirect.com.
6. Research, R., (2015). *Food Industry Sustainability Strategies and Practices Report*. [online] Prnewswire.com. Available at: https://www.prnewswire.com/news-releases/food-industry-sustainability-strategies-and-practices-report–293536091.html [Accessed 17 Mar. 2018].
7. RNR Market Research. *Purchase – RnR Market Research*. [online] Rnrmarketresearch.com. Available at: http://www.rnrmarketresearch.com/contacts/purchase?rname=249114 [Accessed 17 Mar. 2018].

CHAPTER 2

AGRICULTURE

"Agriculture is our wisest pursuit, because it will in the end contribute most to real wealth, good morals, and happiness."

—Thomas Jefferson

CONTENTS

The objective of sustainable agriculture is to fulfill all the needs of the population in terms of food and textile, etc. in the present without compromising the ability of future generations to meet their own needs. Experts in sustainable agriculture search for methods to integrate three main objectives into their work: a healthy environment, economic profitability, and social and economic equity. Each individual involved in the food system whether producers, food processors, distributors, retailers, consumers, and waste managers, etc. can play a significant role in confirming a sustainable agricultural system. This chapter gives a brief overview of agricultural development for sustainable systems.

2.1. INTRODUCTION

The largest source of environmental influence on food and beverage industry is from agriculture. Agricultural production is valued to be accountable for about one-third of the human-induced greenhouse gas (GHG) emissions and about half of the food supply's GHG emissions. It is assessed in the U.S. that 70% of river and stream pollution is produced by agriculture from chemicals, silt, and animal waste. Additional, the largest user of water in the whole world is agriculture, and only 45% of the irrigation water is effectively used.

2.2. DEVELOPMENTS IN WHEAT

One of the most consumed cereals for human is wheat. Wheat hast a global annual production of nearly 600 million tons (Singh and Trethowan, 2007). 50% of the developing country's production is wheat, and much of the genetic growth can be credited to improving variation to a range of water-limiting situations through shuttle upbringing performed by the International Center for Improvement of Maize and Wheat in Mexico. Spring wheat can be grown from the equator to as far as 60° north and south latitudes and from sea level to over 3,000 m elevation. Winter wheat offers another type of adaptation to low winter temperatures and after passing through a vernalization phase can sprout early in spring and produce high yields by midsummer (Figure 2.1).

Figure 2.1: Wheat farm.

Source: https://www.britannica.com/plant/wheat

Together, a wide range of ecoregions adapt to all these wheat varieties and have shown to be particularly drought tolerant. Many of the maximum yields have been reached in Europe under favorable rainfall conditions, while making use of high levels of fertilizer and growth watchdogs to avoid excessive vegetative growth and lodging or falling over before harvest. These structures seem to be sustainable, till the time of the supply of fossil fuels, fertilizers, and chemicals required to produce them are obtainable. To conclude, the sustainability will rely on environmental rules and in what way well human beings are able to use these efforts without creating unnecessary nitrate or chemical residue loads on the environment that are harmful to humans and other species.

The two regions where yields have improved in an exciting technique are northwest Mexico where several of these new wheat varieties were developed, and in the Punjab of India and Pakistan where water has been available and double cropping with a summer cereal or pulse crop has been possible. This type of systems is sustainable till the time fertility can be preserved, subject to the similar restrictions defined for the cereal system in Europe and the obtainability of gradually scarce irrigation water. Water tables have been dropping by way of at least 1 m/year in the scenario of the Punjab due to rigorous use of tube wells and irrigation for both winter and summer crops. At this rate, water rapidly grow into too costly to pump for agriculture.

A critical factor is competition for water from several other sectors. It necessitates around 1,000 tons of water to produce a ton of grain (Gleick, 2000). Together with the present irregular increase in several cereal grain prices, their economic productivity per unit of water is far under that of other industries, and it is additionally not possible to contest with communities for water required for public supplies. The real benefit of agriculture in use of water is the prospective to capture rainfall over a widespread area, accumulate this in the soil profile, and use it to grow crops. The minute that resource is distilled in a stream, reservoir, or groundwater aquifer, it is additionally appreciated to other sectors of society. Even leisure usages and keeping habitat for wildlife species have complex values for some human societies, at least with the present suitable levels of food production in record areas of the world. This could transform as food develops to be scarce and water is required to help source this basic human necessity.

2.3. DEVELOPMENTS IN RICE

Another important cereal grain is rice, also with annual global production over 400 million tons (Virmani and Ilyas Ahmed, 2007), and 90% grown in Asia. The crop can be produced from 35◦ south in Australia to 50◦ north in Mongolia, and from sea level to over 3,000 m elevation. The highest level of productivity is in Egypt and Australia, with average yields over 9 tons/ha. In the concluding half of the past century, the area of rice growing has increased almost 1.8 times in size, while yields per hectare have increased more than to almost double, consequential to four times increase in global production (Figure 2.2).

Figure 2.2: Rice crop.

Source: https://www.farmcrowdy.com/farm/rice-farm/

Most significant technical developments have contained within breeding diversities that are unresponsive to photoperiod, and thus can be planted in any month of the year where water and temperature conditions are favorable; semi-dwarf varieties that answer to fertilizer with more grain before vegetative development; shorter maturity varieties that permit two or three crops per year if irrigation is obtainable; and chemical nutrient and weed management to back the extremely extractive practices related with high yields and multiple crops per year.

In addition to restrictions on water, the most troublesome subjects to begin over the last two decades with rice has been yield decline in Asia. Even though a number of concepts have been planned, it now gives the impression that nitrogen obtainability at the appropriate times in the crop's life cycle is one of the principal issues (Doberman et al., 2000). Research on this serious issue remains, as rice makes for a significant component of the diet throughout Asia. There are growing number of apprehensions, including the option of sub-detectable things of soil pathogens, multifaceted queries of nutrient obtainability in a nonstop cultivation pattern of the same crop, and other potential soil nutrient reasons for the decline. It is vital that this issue be resolved for the comfort of millions of people in Asia. The possibility for resolving the test through crop rotation is a clear route to take, yet the appropriateness of these lands for rice cultivation and the ongoing demand for this widespread food crop are overruling explanations to discover methods to make incessant cultivation sustainable, as problematic as this may be geographically.

2.4. ADVANCES IN MAIZE

One of the most important cereal crop in terms of total production annually is maize. It is approximately 700 million tons annually. This is a chief cereal in much of Africa and Southeast and East Asia, far from its source in Guatemala and southern Mexico. As the primary crossbreed crop to obtain major consideration from plant breeders, maize has developed to be a model for plant genetic development and the most vital cereal grain in the United States and several other temperate countries. Additionally, for the choice for crop yield, initial efforts attentive on increasing protein and oil focuses in the grain with the expectation that this would not reduce grain yields (Johnson, 2007). Growth of inherited lines of maize and their heterozygous crosses turn out to be the normal for study of population genetics, testing methods, and more lately marker-assisted choice and other microbial technique (Figure 2.3).

The hybrid maize industry in the United States has been formed on the basis of double-cross maize hybrids (four inbred parents) and then single-cross hybrids (two inbred parents) and a model for other countries seeking high yields. This resulted in a wide adaptation of new genetic combinations. New plant breeders with maize presented population progress and other varieties that could be produced and their seed preserved by farmers in addition to the hybrids based on inbred lines. This idea was developed on information of the pollination habits of maize; 95% of the pollen fertilizing a given ear on a plant comes from another plant, thus it can be said that the crop is 95% outcrossing.

Figure 2.3: Maize crop.

Source: https://www.britannica.com/plant/corn-plant

It can be said that 95% of the seed that arises from a choice in the field is a consequent of crosses with some other series of plants in that field or nearby. By combining a variety of plants with alike features, together with grain color and crop maturity, one can harvest seed from the "best" plants in the field and guarantee that the large majorities are hybrids between two parents in that same field. By starting with a fairly varied and extremely creative change (population or synthetic variety), it is likely for the farmer to choose a better variety that will be even better for his or her particular farm conditions.

It is vital to do this choice of plants in the field, selecting those individuals that stand up well, have insect and pathogen tolerance, and are well adapted to the cropping system.

A comparatively innovative development in crop advance has been the overview of transgenic hybrids of maize, rather erroneously called GMOs, or genetically modified organisms. This is a contradiction because all of our farm crops have been genetically transformed since the first farmers chose plants with larger seeds or good food quality to increase and plant near their homes.

2.5. POSTHARVEST LOSS

The Food and Agriculture Organization (FAO) has since a long time ago perceived the potential for enhancing food accessibility that can accompany decreases in postharvest losses (FAO, 1981). Postharvest losses have been evaluated to be around 21% of the aggregate food in our current supply chain (Niranjan and Shilton, 1994). Postharvest losses originate from each phase after the food has been removed from the plant (sea or creature), from harvesting, handling, storage, and transport. The reasons and measures of losses fluctuate extraordinarily and rely upon the yield or food and its area. At that point it is imperative to plan appropriately for the harvest, planting what is required (with some overage), harvesting in the right conditions, and harvesting at the correct time. Harvesting ought to likewise have insignificant physical effects on the product to prevent accelerated physiological deterioration (Figure 2.4).

Figure 2.4: Postharvest loss.

Source: https://emkambo.wordpress.com/2015/09/14/the-curse-of-pre-and-post-harvest-losses-in-african-agriculture/

Additionally, control measures after the field that lessen losses are appropriate storage conditions (temperature and moistness), which for a few foods can mean immediate cooling. It is accepted that reducing high postharvest losses requires technological developments.

Despite what might be expected, access to the right methods intends to prevent the losses is the biggest obstacle for loss reduction. Restricted access to the suitable methods for loss control can cause noteworthy setbacks. For instance, postharvest losses in small-scale fisheries can be among the most elevated for all products (UN Atlas of the Oceans, http://www.oceansatlas.org/html/moreinfo.jsp, got to June 11, 2008) frequently because of restricted refrigeration and freezer facilities, particularly onboard the fishing boats.

2.6. ORGANIC FARMING

Organic farming is regularly considered as an economical reasonable method for sustainable food production. There has been a yearly increment of more than 20% in the organic food market in the United States for as long as two decades (Om Organics, 2008). This is a critical accomplishment and one of the only major growth areas in an industry that generally considers food markets as inelastic, only growing with population and as subsequent demand increases. Organic farmers manage nutrients without utilization of substance manure, utilizing a mix of crop rotation of species with various nutrient needs, application of animal manure and/or compost, soil-building cover crops in the sequence, and calculating a careful nutrient budget to assess crop removal as well as potential for building soil fertility over time (Figure 2.5).

Good managers can lessen expenses of purchasing and applying excess nutrients, and in the meantime avoid from adding to environmental issues because of nutrient misfortune from farm fields. Organic farms are known for utilization of manure and compost and continuous reconciliation of crops and creatures on same farms to make nutrient cycling more proficient (USDA, 2007). In spite of the fact that there might be over the top cultivation and resulting soil erosion on some organic farms, this way agriculture does reduce chemical load in the environment and engenders efficient use of on-farm, renewable production resources.

Figure 2.5: Organic farming.

Source: https://www.organicfacts.net/organic-farming-india.html

Another effect of innovation has been an expanding industrialization of organic agriculture. The engaging quality of the organic segment of the food business is progressively perceived by those in the worldwide agriculture and food parts, and today over half portion of all organic food is marketed through substantial corporate grocery stores instead of small, privately owned, specific organic food stores. There is worry among a large number of the organizers of the organic food movement that a significant part of the social purpose of the original concept is lost, and they are searching for a more prohibitive sort of certification that would emphasize more than production strategies: reasonable treatment of farm work, focus on nearby foods, and distribution of advantages to a more extensive gathering of citizens. As opposed to the possibilities detailed by most creators, there is still open deliberation about potential worldwide production from organic agriculture (Sustainable Food News, 2007).

The FAO expressed that organic agriculture ought to be utilized and advanced for its healthy and nutritious value and additionally the growing income it is accommodating for developing and developing nations; in any case, with current yields and land use, it will be unable to feed the 6 billion individuals on the planet today and the possibly 9 billion in 2050 without judicious utilization of compound fertilizers (Sustainable Food News, 2007). We challenge that conclusion, in light of individual perceptions, as the best organic farmers in the Midwest reliably produce crop yields as high if not higher than regional averages.

2.7. LEGISLATION AND SUPPORTS

Technologies in agriculture, particularly in the United States and the European Union, have been greatly fruitful in raising efficiency and expanding creation far beyond internal needs and nearby markets. This has prompted significant exports of cereals from these places and development and later consolidation of grain market into a few major companies. Although frequently held up as a model for the success of free-market economies and advertised as an industry that has profited farmers in the North and in addition food-deficient nations in the South, in fact, the agricultural export industry has conveyed balanced advantages to the bigger farm operations and to those providing modern agriculture. This incorporates organizations providing inputs, commodity merchants, and exporters. While profiting these bigger operations, the farming export industry has frequently suppressed fruitful production and skewed markets in numerous developing nations. The North American Free Trade Agreement is the most recent case of this movement of the free market agricultural economy.

A long way from making a level playing field, the enactment and supports in the two noteworthy food export nations in North American have supported large-scale operations and funded trades in various ways, while small farmers in Mexico have been forced out of cultivation. A standout amongst the most prominent liberal economists in agriculture, and consultant to presidents from both major parties, is Willard Cochrane from Minnesota. As per Cochrane, "long-term attempts to stabilize production and farm incomes through farm programs in the United States have surely done some good by providing subsidies for maintaining prices during hard times and promoting export of food grains" (Cochrane, 2003). However, they have failed to give long-term dependability, and the outcome throughout the 70 years has been a continued consolidation of possession, exit of farmers and farm families, and fall of rural communities.

Cochrane additionally infers that vast territorial cooperative projects, for example, the North American Free Trade Agreement have added to more noteworthy, instead of lesser, imbalances in incomes and achievement, and are particularly damaging for little and family farms in the United States and in the other two partner nations. Another regarded agricultural economist in the land grant framework, John Ikerd from the University of Missouri, keeps up that the way toward long-term security in the United States food system is through sustainable agriculture (Ikerd, 2006).

Keeping in mind the end goal to comprehend the negative and sudden ecological outcomes of the current industrial model of agriculture, it is fundamental to decrease the contamination of waterways and aquifers from pesticide residues and chemical fertilizer nutrients. These residues originate from more than 1.1 billion pounds of pesticide applied every year (Kiely et al., 2004) and more than 12 million tons of nitrogen applied yearly in the United States (ERS, 2007).

The residue in the environment from these chemicals are indicators of the decrease in biological sustainability of present food production system. As Ikerd attests, acknowledgment of the ecological effects of conventional agriculture has prompted more noteworthy scrutiny of the economic and social sustainability of these same systems. Consolidation has halfway been a consequence of farm support programs (Ikerd, 2008), since payments have been coupled to production, enabling the bigger administrators to gain more capital, which is then put into land purchase.

With little net revenues on conventional commodity crops, the regular knowledge in the western Corn Belt is that a family should cultivate no less than 1,000 acres of land to earn enough net income to help a sufficient way of life. Similarly, as with any standard way of thinking, this speaks to a normal farm size —some farm families add value to diversify their crops and animal enterprises reduce costs by making use of primarily internal resources from the farm for maintaining soil nutrition, manage pests through rotation and diversity, and direct market their products often claim that a much smaller farm is adequate. There is a developing interest for the United States and a strong initiative in the European Union in perceiving the significance of multifunctional rural landscapes, particularly as they provide a range of ecosystem services to the larger society.

One of the administrations that is already perceived economically and traded on the future markets is carbon credits. There are likewise government programs (United States) and territorial programs (European Union) that reward preservation related practices with yearly subsidies. One of the clearest mandates on horticultural research in recent years in the United States that supports this new course was the complete report of the National

Research Council on priorities for research in future. The prestigious board that assembled this report evidently distinguished the significance of a multifunctional agriculture and rural group development as the establishment of viability for the rural sector in the United States (National Research Council, 2003), and declared that agricultural landscapes are vital for more food production. In this manner, they recognize the significance of families, groups, and biological services.

However, we perceive that it will be a very long time before a research establishment, as large and complex as that in the United States, one driven essentially by agricultural production and upheld emphatically by input and grain companies, will roll out any real improvement in direction. It is obvious that we do require a more extensive way to deal with advertising in a mind-boggling and greener future that incorporates more environmental concern by purchasers and investors. Although the high-innovation way to deal with agriculture has brought about fast increments in efficiency and production, and one result is an expanded accessibility of food to many, there have been some unforeseen and negative outcomes (Horrigan et al., 2002). The attention on crop yields and maximizing profits as a solitary technique has prompted monetary, natural, and social issues (Shrestha and Clements, 2003). Keoleian and Heller (2003) summarized these outcomes.

They found that quick change of prime farmland to urban advancement prompted less steady and less arable land being utilized for agriculture resulting in increased erosion, and water system leading to depletion of topsoil exceeding regeneration and rate of groundwater withdrawal exceeding recharge in major agricultural districts. Losses to pests are increasing, despite use of chemicals. With a 10-fold increment in insecticide use from 1945 to 1989, there was a simultaneous increase in losses from insects from 7 to 13% (Pimentel et al., 1991). More recent estimates recommend that there is a loss 37% of all crops, internationally, because of pests (insects, pathogens, and weeds), regardless of enormous utilizations of pesticides (Pimentel and Pimentel, 2008). There is a lessening in hereditary diversity, since today just 10–20 crops give 80–90% of the world's calories (Brown, 1981). Such absence of biodiversity makes the supply more vulnerable to pests and disease, leading to declining economic conditions for farmers, particularly as production moves from smaller farms to larger, industrial farms.

This is on account of the cost of crops is low, yet investment is high. This implies that progressively just the bigger operations can make a decent living. In this way, we have a complex worldwide food situation, where a lot of food are lost to pests, expenses of data sources are expanding quickly, there is rivalry between production for food and fuel, and extraordinary imbalances exist between the North and South. While numerous on the planet are undernourished, various nations in the North are seeing overconsumption and weight as developing wellbeing challenges. These subjects are extended, as both social and monetary effects of the present food system. There is an overwhelming dependence on non-renewable energy sources bringing about an awkwardness in vitality info and yield that further weights the framework and condition.

On a worldwide scale, agriculture fundamentally adds to greenhouse gas (GHG) discharges, delivering in the area of 17 and 32% of all worldwide human-induced GHG outflows (Bellarby et al., 2008). This incorporates land and livestock direct emissions, non-renewable energy source use in cultivate tasks, production of agrochemicals, and energy expenses of the general sustenance system. The biggest contributors are change of land to agriculture and direct discharges from land and animals. For instance, abundance utilization of composts discharges nitrous oxide, a GHG, and fertilizers release of nitrous oxide is the single biggest contributor of GHG in agriculture (Bellarby et al., 2008).

This does exclude GHG contribution for the production of fertilizers. Livestock is the biggest user of land, utilizing 33% of the world's arable land due to the move far from grazing to the development of livestock feed crops (Compassion in World Farming, 2008). This has caused major deforestation and other local vegetation destruction. Such change is a net arrival of carbon, given that local vegetation, for example, trees, stores more carbon than crops like soy (Bellarby et al., 2008).

Bellarby et al. (2008) propose that the huge environmental effects of agricultural production can be moderated with various methodologies. These emphases on moving farming practices to alternatives that give carbon sequestration as opposed to emission, for example, enhanced cropland administration, (e.g., avoiding bare fallow/using cover crops and appropriate fertilizer use), grazing land administration, and rebuilding of natural soils as carbon sinks. Further, since meat creation is wasteful in its conveyance of vitality to the human evolved way of life and in the meantime is a noteworthy wellspring of GHG outflows, a lessening of meat production and utilization could give major improvements. It has been recommended that a lessening in the measure of the livestock industry is the most straightforward, rapid, and likely the main successful technique for cutting GHGs from animal production to the degree that is important to restrain the future increment in global warming (Compassion in World Farming, 2008). The difficulties in agriculture and the worldwide food system are numerous and complex.

2.8. CONCLUSION

The quality and safety of food will remain a top priority for food processors, particularly due to growing numbers of authenticity, adulteration, and safety problems in food products. Though, environmental sustainability is turning out to be of progressively important to companies. The co-management of safety, quality, and sustainability issues will guarantee that quality and safety are continued while confirming enhanced sustainability. The key challenge in realizing this goal will be allowing companies in making informed decisions. Food processors should not only evaluate the effect of processing but also the broader supply and distribution chains, and end of life of the product or packaging. The use of several sustainability valuation procedures should contribute in achieving this goal.

REFERENCES

1. Baldwin, C., (2010). *Principles of Sustainability for Food Products*. [online] Food manufacturing. Available at: https://www.foodmanufacturing.com/article/2010/08/principles-sustainability-food-products [Accessed 17 Mar. 2018].
2. Future Food, (2016). *The Future of Food: Ten Trends in Food Sustainability – Future Food*. [online] Available at: http://futurefood.network/the-future-of-food-ten-trends-in-food-sustainability/ [Accessed 17 Mar. 2018].
3. Intertek.com. *Sustainability Solutions for the Food Industry*. [online] Available at: http://www.intertek.com/sustainability/food-drink-industry/ [Accessed 17 Mar. 2018].
4. Murphy, F., McDonnell, K., & Fagan, C. (2014) *Sustainability and Environmental Issues in Food Processing*. [online] Nfscfaculty.tamu.edu. Available at: http://nfscfaculty.tamu.edu/talcott/courses/FSTC311/Textbook/9-Chapter%209%20Sustainability%20and%20Environment.pdf [Accessed 17 Mar. 2018].
5. Pretty, J., Morison, J., & Hine, R. (2001)*Reducing Food Poverty by Increasing Agricultural Sustainability in Developing Countries*. www.sciencedirect.com.
6. Research, R., (2015). *Food Industry Sustainability Strategies and Practices Report*. [online] Prnewswire.com. Available at: https://www.prnewswire.com/news-releases/food-industry-sustainability-strategies-and-practices-report–293536091.html [Accessed 17 Mar. 2018].
7. RNR Market Research. *Purchase – RnR Market Research*. [online] Rnrmarketresearch.com. Available at: http://www.rnrmarketresearch.com/contacts/purchase?rname=249114 [Accessed 17 Mar. 2018].

CHAPTER 3

QUALITY, SAFETY, AND SUSTAINABILITY IN FOOD INDUSTRY

"It is impossible to improve any process until it is standardized. If the process is shifting from here to there, then any improvement will just be one more variation that is occasionally used and mostly ignored. One must standardize, and thus stabilize the process, before continuous improvement can be made."

—Masaaki Imai

CONTENTS

In this chapter, we will study the different processes applied to maintain food quality, food safety and sustainability. This chapter starts with the discussion on the HACCP system, which is a profitable method for ensuring the safety of food. HACCP is an internationally accepted method and an effective tool which deals with safety hazards that arises in the food production process. Here, we will discuss in detail about the risk management by the HACCP system, the importance of the HACCP system for a sustainable development, and risk factor in the integration of the HACCP for a sustainable development policy. The next topic discussed will be the food distribution chain, its types and temperature coordinated distribution. Next, we will discuss about food specific issues in distribution management. In the end, distribution network design, as well as packaging and labeling of food products, are discussed.

3.1. INTRODUCTION

A food system is a system which involves several distinct aspects; a central aspect is to transport food from the place of production to where people can buy and eat the food and from there to disposal (Neff et al., 2009). The various levels which are included in food system are production, processing, distribution, preparation, marketing, access, consumption, and disposal. All these processes require resources such as businesses, communities, farms, interventions, people, policies, and politics. All steps need to be coordinated for a company to make progress (Miljöstyrningsrådet, 2007). A food safety standard helps to coordinate this through guidance and instances and see that the distribution networks are running according to present principles as well as regulations. There is a higher knowledge of risk analysis in food production when the demand of food export increases. A risk analysis document needs to be continuously updated and to work towards it there are many methods, principles as well as standards. A risk is defined as the likelihood of an event to occur and the consequences of it (ibid.). Risk management is all about methodically assessing as well as handling risks associated with the operation. According to Swedish Board of Agriculture, "Sweden has increased the export of commodities from agriculture and food industry in recent years" (Jordbruksverket, 2011).

The food industry sector has become aware of all the risks which are noticed by the type of installation as well as how it is affecting the population health, number of questions related to the management system have been raised which provides reliable information on the safety and quality of the products which are sent to the consumers for their day-to-day use. It is indispensable to check the quality of food at every level taking proper measures into consideration, as there is very high possibility that food can be contaminated in diverse ways as well as at diverse levels which can lead to several diseases, such as, for instance, abortions, injuries, nervous, and digestive disorders, vomiting, fever, choking.), the cause of disease can range from normal fever and can even cause death. Every company who manufactures sells or transports food is facing the same the risks of contamination. There is a risk at every level which can occur at any link in the chain representing the series of the phases through which products move from their entry to the exit of the company. The food security management system consists of acting as a barrier for the faulty components and the defective end products or goods. The classical phase of the management of the safety and quality of food constitute of a reception control for the purchased materials, inspections during the manufacture and verification of the finished products. In the food sector the Codex Alimentarius Commission creates the FAO, and the WHO which was founded in 1962. Some other regulatory references such as ISO 9000, 9001, 9004 and ISO 22000 all of them have a joint objective which is to control risks in this type of sectors. With the implementation of HACCP (Hazard Analysis and Critical Control Point) system the food security has undergone a considerable development, HACCP is a methodical approach whose objective is to find, estimate, and control the physical, chemical as well as biological risks along the whole production process, and therefore it has become the tool of control of the food hazards. "HACCP constitutes a reference base to the food safety, it provides a systematic method for analyzing the food processes, identifying potential risks and specifying their critical control point (CCP), to avoid the distribution of unsafe food to consumers."

Since food conditions are not in their favor most of the times, large part of humanity faces serious threats. Therefore, there is a huge demand to find methods which permanently ensure food safety and preserve the natural capital which conditions it after confronting with the threats faced in food sector. Mostly, it is very difficult to take up such decisions due to various reasons, some of these reasons include economic problems, legal issues, political considerations, and social values.

But science is still playing the most crucial part in decision-making for food risks assessment. Due to these reasons the management aspect of food risks is measured as vital in the policy of food safety based on rigorous scientific standards.

3.2. HACCP SYSTEM

By the involvement of Hazard Analysis and Critical Control Point (HACCP) tool, the innovative integration of a sustainable development approach will be made by progressive actions using the results of a management. The HACCP tool's opportunities will:

- Allow risk management by using control measures that have been executed in health, food safety, environment, as well as working conditions;
- Help in elaborating a common and progressive working document that deals with hygienic safety problems while suggesting a synchronized method. Some of the original methodological approach which will demonstrate two research axes are:

1. The first approach is centered on demonstrating the interest of the HACCP system implementation so as to provide the numerous companies with an effective tool which measures as well as masters the risks which are threatening food healthiness;
2. The second part of the work helps in establishing an integration framework of this self-control system in a policy of sustainable development. The aim of which is to ensure the durability of its activities by preventing the progress of serious pathologies with nutritional constituent, and also to find an innovation tool for the food-processing industry. The base of integration of the approach revolves around these points (Figure 3.1):

a. The food industry on the way of durability;
b. The Risk management by the HACCP system;
c. The importance of the HACCP system for a sustainable development.

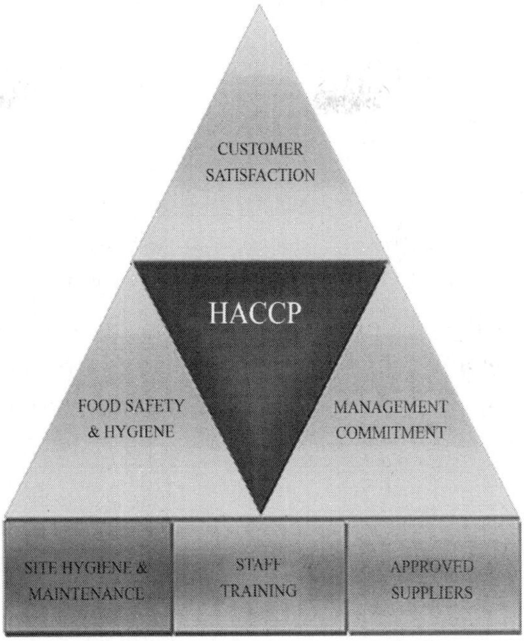

Figure 3.1: The HACCP system.

Source: http://www.millersofspeyside.co.uk/quality-assurance.php

3.2.1. The Risk Management by the HACCP System

HACCP (Hazard Analysis and Critical Control Point) is so far recognized as a profitable method for ensuring the safety of food. HACCP is an internationally accepted method and an effective tool which deals with safety hazards that arises in the food production process. In order to prevent the occurrence of food safety hazards to final customers, the use of its seven application has become mandatory prerequisites in the food worldwide production chain. It has been advised to implement the full HACCP in all food businesses in order to strengthen public health security. The planned principles are different from those of the Codex and there is no reference of the need for normal HACCP plans. In order to reduce the risks to suitable levels, HACCP identifies risks in the production processes that can lead to unsafe products, and designs measurements.

HACCP is designed for application in all links of the food chain, and there are seven principles around which HACCP revolves:

1. Analyze hazard (biological, chemical or physical);
2. Identify critical control points (these are points in a food production at which the potential hazard can be controlled or eliminated);
3. Establish protective actions with critical parameters for each control points;
4. Establish measures to monitor the critical control points;
5. Establish curative actions to be taken when monitoring shows that a serious limit has not been met;
6. Establish procedures to verify that the system is working properly;
7. Establish effective recordkeeping to document the HACCP system.

In order to manage food risks, critical control point huge system step has been made with the application of HACCP implementation which helps not only to recognize the risks and control them and especially strengthen the self-control of the food-processing companies. There is a very specific HACCP system for each company where each case establishes an exact study. Thus, implementation of this system depends on the company's decision. There are different detailed assistance tools for the HACCP system. For example, for the determination of the CCP, we use the tool "tree of decision" whose elaboration is made thanks to a set of questions related to every risk and to every stage of the process, including the reception and the processing of raw materials. For instance, tree of decision tool is used for determining the CCP.

3.2.2. The Importance of the HACCP System for a Sustainable Development

The food safety management policy is a method to sustainable development whose major goal is to gather data based on scientific criteria while respecting the vital values associated with food safety. Codex Alimentarius largely takes into account the vital values related to food safety, which is composed of food standards and applied codes giving guidelines and references on food security as well as on food security management systems on HACCP system model. HACCP system is a tool intended to measure risks and to establish systems of control centered on prevention instead of calling upon events and actions of a posteriori control of the end product numerous works were directed in this domain to show the importance of the integration of the sustainable development in management practices. In practice, and for reasons of efficiency, "food safety should integrate, in its management," "requirements for sustainable development whose objective is to offer a product of quality while respecting the principle of balance between economic growth," "social development and environmental protection."

3.2.3. Risk factor in the Integration of the HACCP for a Sustainable Development Policy

The most propagated risk after the application of the step HACCP system is the presence of pathogenic bacteria which can have destructive effects on the consumers. There are certain parameters which are determined by the application of this approach and it is essential to act (CCP) and the control such measures, thus assuring the productivity of preventive measures and reducing the necessity on the reviews and checks of finished products.

From the primary producer up to the consumer (on the whole life cycle of products) the HACCP system can be applied all along the food chain in order to improve the salubrious safety of the food. On the other hand, the HACCP system allows a better use of the energetic as well as economic resources. The HACCP system guarantees a long-lasting consumption by designing appropriate products and services that necessitate less energy and resources, and is not presenting any harm or threat. Given below is the figure which defines a plan of grouping together the key dangers and risks in the sustainable development that must be considered by the industries (Figure 3.2).

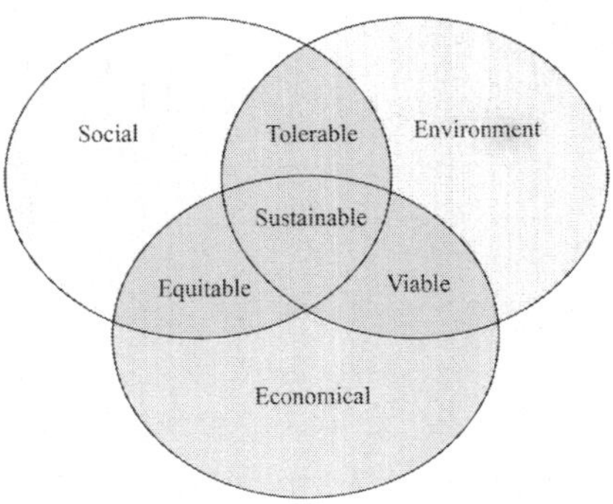

Figure 3.2: Main stakes of the sustainable development in food industry.

Source: https://file.scirp.org/pdf/JSSM_2014043010065191.pdf

In order to set up an integrated method, the overall objective of the above figure is to identify the economical, environmental, and social aspects in the sustainability field that will support Businesses to develop a plan of integration with regard to their actions. The objective of this figure is to bring in to light the health factor as a pillar in this approach of sustainable development while integrating into the interface the other themes, namely the social, environmental as well as economical factors.

1. **Social**

- Hygienic quality by traceability actions and the outcomes of the HACCP;
- Training of professional workforce;
- Communication as well as information about the obtained outcomes;
- Assure food quality through quality labeling and classifying;
- Use of ergonomics and studies to upgrade the working conditions about the medium hygiene;
- Be economically responsible and accountable of buying for products from convenience stores.

2. **Environment**

- Decrease water and energy consumption by professional staff;
- Use cleaning equipment's with weak and energy consumption;
- Decrease raw material consumption by the amelioration of food consumption systems and fight against all kinds of waste;
- Limit the costs by the actions of energy and raw material saving.

3. **Economical**

- Be economically responsible of buying for products from convenience stores;
- Establish confidence between supplier, customers, and consumers.

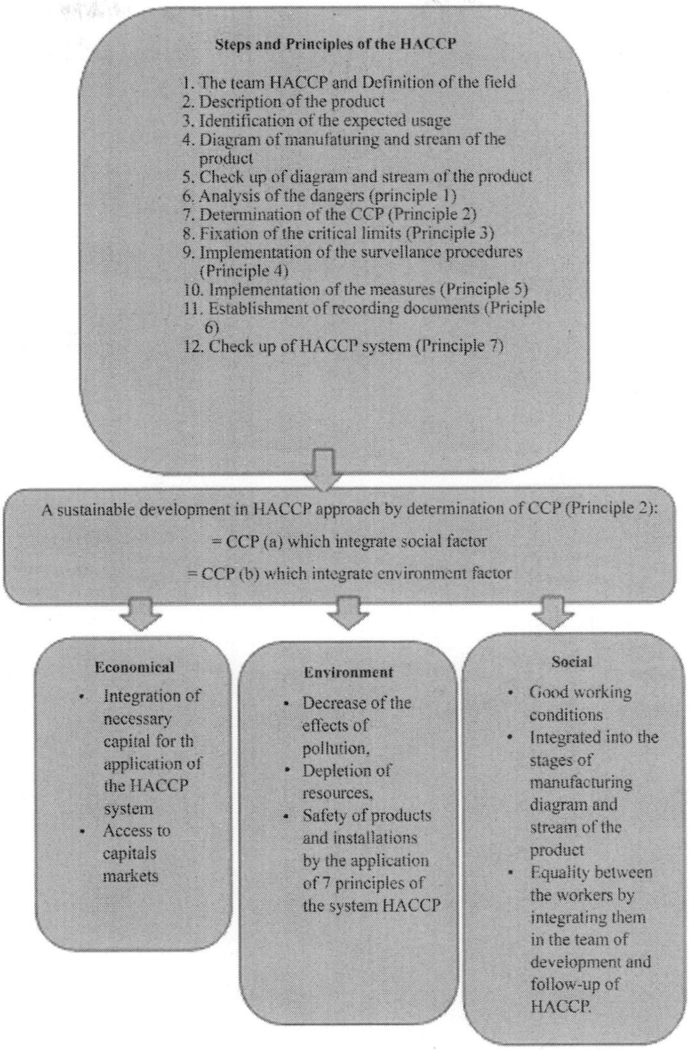

Figure: Steps and principles of HACCP.

3.3. FOOD DISTRIBUTION CHAIN

Food distribution is very different from the distribution of other products. Until final consumption all the way, food products show constant quality changes throughout the supply chain. Before food reaches to consumers it undergoes many quality, health, and safety checks. Due to the occurrence of food scares, there is a need to repeatedly and vigorously check the quality and safety of food. For example, if food is contaminated it may lead to severe illness or even death, some of the diseases are caused by the presence of salmonella in chicken or by cows infected with BSE (Figure 3.3).

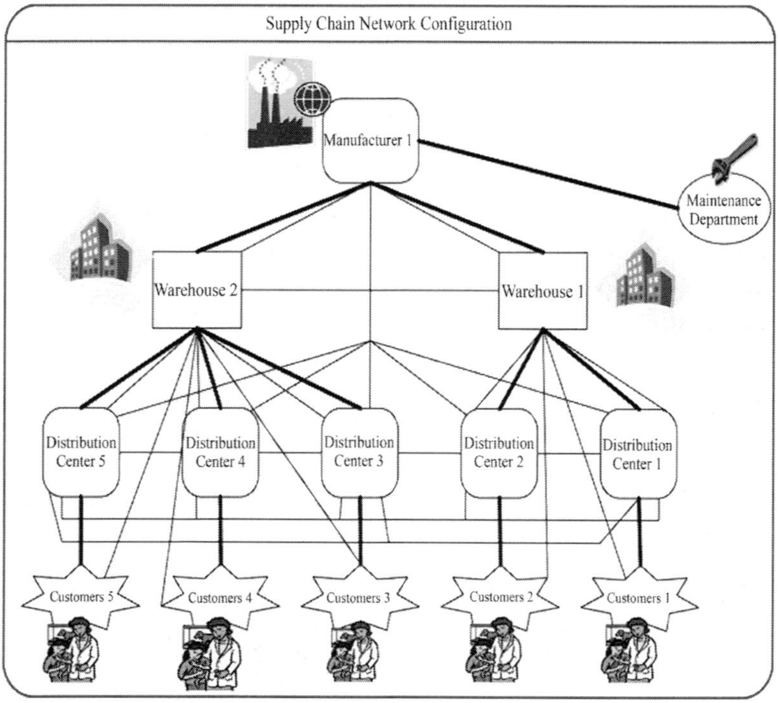

Figure 3.3: Supply chain network configuration.

Source: http://cmuscm.blogspot.in/2014/02/cost-of-weak-relationships-in-supply.html

In the current globalized economy, the distribution of food is the main point of discussion in society as well as in academic literature and the food industry has been at the lead in developments associated with sustainability.

Food distribution management is a challenging area that has only recently began to receive more attention in the operations management literature the reason behind this stagnant growth in this sector is the limited shelf lives of food products, time windows for delivering the products, possible interaction effects between products, requirements with regard to temperature and humidity, high customer expectations, and low-profit margins.

Some of the major challenges that are faced by the food industry are:

1. Food safety;
2. Food quality;
3. Food sustainability.

3.3.1. Food Distribution System

The first step in the food supply chains is the manufacturing stage, as well as foodservice or retail activities in which forward integration method is adopted, i.e., how the final product made from agricultural producers reaches to consumers. As stated by Rushton et al. in 2006, "Distribution management normally refers to the physical flows and storage of products from the final production point to the customer or end user." The food distribution system does not involve the initial stage of the supply chain, that is, how manufacturers get the raw-material from agricultural producers. Food manufacturers obtain agricultural raw materials and process those before further distribution. Food processing is a broad sense that ranges from normal packaging of fresh produce to extensive cooking or preservation processes. Before the production and distribution of crops the part of the supply chain was focused on other reviews. For most of the food distribution systems, the important characteristic is temperature control (Figure 3.4).

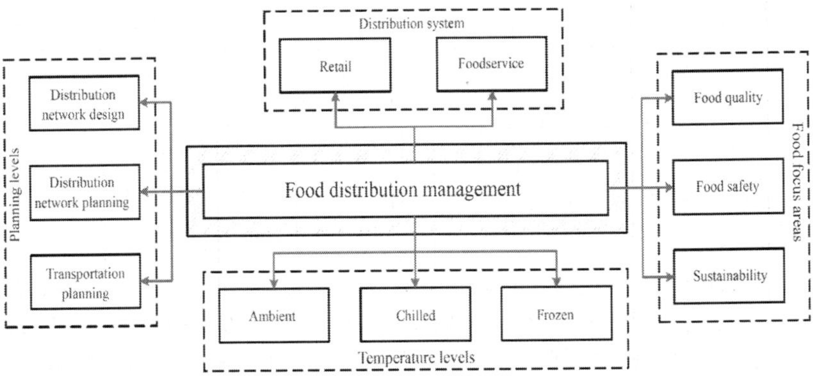

Figure 3.4: Food distribution management.

Source:https://www.sciencedirect.com/science/article/pii/S2210539517300901

Temperature control is important for controlling food quality and food safety for the extensive variety of products. However, it also leads to additional energy consumption. Temperature-controlled distribution is associated with all three specific aspects of foods, that is food quality, food safety and food sustainability.

3.3.2. Types of Food Distribution Chains

After the manufacturing stage, comes the wholesalers, foodservice businesses and retailers. They are the three main actors who distribute the food products through the supply chain to the final consumers. In 2001, Eastham et al. distinguished retail as well as foodservice from a consumer viewpoint, chiefly by mentioning to the consumption site, which is, respectively, inside or outside a hospitality operation. Due to the increase of ready-to-eat meals sold in retailing and the establishment of restaurants within retail stores, they acknowledge that this boundary is increasingly fuzzy. Both industries are still quite different regarding the whole chain in terms of, e.g., outlet dispersion, supply volumes and the use of technological systems like Electronic Data Interchange (EDI) or Electronic Point of Sale (EPoS).

For the retail industry, the main actors ranging from smallest to largest are- specialist shops, smaller convenience retailers, and large supermarket chains. As said by Dobson in 2001, "The industry has seen significant consolidation and concentration, which has led to domination of the market by large retailers." Huddleston et al. in 2009 stated that "Specialist shops often gain their competitive advantage due to a deep product assortment and a focus on high-quality products." However, a wider range of products are offered by many conventional large retailers. In recent years, as observed by Boyer and Hult in 2005, the online retailing has increased to manifolds which often leads to diverse delivery channels, such as direct shipment from producers to consumers. There is special attention in the distribution systems of such online retailers to, for example, "pricing schemes based on delivery time windows in relation to expected routing costs" (Campbell and Savelsbergh, 2006), or "the creation of special distribution centers for the fulfillment of internet customer orders" (De Koster, 2002).

As specified by Taylor in 2008, "The main actors (or customers) in the foodservice industry are restaurants, cafes, takeaways, street vendors, hospitals, schools, prisons, residential homes, hotels, and other premises where food is produced for immediate consumption." SMEs (small- and medium-sized enterprises) often small-scale production units that take care of their own distribution network dominates the foodservice industry. The industry does, however, experience a trend towards industrial scale food production (Engelund et al., 2009). Due to trends in demography and lifestyle, such as the increase in the number of elderly people, the number of one-person households and the rising share of out-of-home dining the consumption of prepared meals is increasing and will continue to grow in the foreseeable future (Buckley et al., 2007). Identifying the manufacturing stage is not straightforward in some foodservice chains, as in some cases the product is often processed in several stages. For instance, agricultural raw materials are first processed at food manufacturers and then further processed in kitchens of foodservice operators.

In retail chains, an increased availability of data has led to industry initiatives like aiming to improve the efficiency in the collaboration between producer, distributor, and retailer, efficient consumer response (ECR), and establishing an efficient product flow. This is often accomplished by forming a pull-distribution system, based on an IT system that relies on EPoS data (Hoffman and Mehra, 2000). The leading party in this effort are the large retailers that dominate the retail market.

For the foodservice industry, as stated by Hill and Scudder 2002, "there are no such large dominating parties, but a similar initiative was started by some industry organizations under the name efficient foodservice response (EFR)." Despite initiatives like ECR and EFR, it should, however, be noted that in reality the integration opportunities between different players in a food distribution system are often partial by product or production characteristics (Van Donk et al., 2008) or obstacles preventing information sharing (Lee and Whang, 2000).

3.3.3. Temperature-Controlled Distribution

Temperature control is an essential factor in the distribution of food products as temperature affects product safety by influencing the growth of potentially harmful bacteria (such as Salmonella and Escherichia coli) and affects product quality by influencing the level of quality degradation. Also, insufficient, or varying temperature control often lead to chemical reactions which ultimately changes the appearance or texture of the product. The shelf life of the food product is also determined by these undesirable changes in the product, which is hence often related to a temperature requirement. Three types of food supply chains can be basically identified in relation to temperature control, these three types of food supply chains are: frozen, chilled, and ambient. A number of different temperatures are used for the frozen and chilled chain. The best suitable temperature for frozen chain is at −18°C, however products such as ice creams, etc. requires an even lower temperature of −25°C. For the chilled chain, temperature may range or vary from 0°C for fresh fish to 15°C for bananas and potatoes (Smith and Sparks, 2004). Lastly, an ambient chain is there which do not need temperature control, for example, products such as canned goods comes under the ambient chain. The exact temperature levels within the frozen as well as chilled chain are significant for the products involved, but a basic classification of frozen, chilled, and ambient is adequate in relation to the discussion of the food delivery management literature.

The main modes of handling products are reflected by this kind of classification in terms of production as well as distribution technologies, related to temperature control and product packaging, (e.g., cooling equipment or insulating packaging material). It also resembles to diverse ways of dealing with quality degradation, which in a frozen state may be nearly still for some products. Several studies investigate heat transfer and microbial growth during transport. A large share of the shelf life must be left for the final consumer for chilled products bought from retailers, as maximum of these products spend a considerable time in home refrigerators, where temperature is often higher than the suggested temperature (James et al., 2008), which leads to rapid degradation in quality.

3.4. DECISION-MAKING PROCESSES IN DISTRIBUTION MANAGEMENT

Managerial decision-making is commonly divided into different levels of decision, mainly relating to the time horizon for these decisions (e.g., Anthony, 1965; Bitran and Tirupati, 1993). This usually leads to the distinction between short-term planning, mid-term planning as well as long-term planning, or alternatively, tactical, strategic, and operational or operative planning. Three distinct planning levels in distribution management are further elaborated in this hierarchical approach:

- "Distribution network design, concerning long-term decisions on the physical distribution structure." For example, this includes the number and sizes of warehouses and cross-docking points, as well as their associated transportation links.
- "Distribution network planning, concerning mid-term distribution planning decisions related to fulfilling demand (or forecasts) on an aggregate level." For example, this includes collective product

flows as well as delivery frequencies.

- "Transportation planning, concerning short-term planning of the distribution of actual customer orders." For example, this includes the loading as well as routing of vehicles.

There are some typical decisions that are mentioned in the list for each of the planning levels. "Planning decisions are typically made based on cost or profit evaluations" (Shapiro, 2007). The table given below summarizes the characteristics of distribution planning on the different hierarchical levels (Table 3.1):

Table 3.1: Characteristics of Distribution Planning on Different Hierarchical Levels (Based on Shapiro, 2007)

	Planning horizon	**Time representation**	**Objective function (s)**	**Frequency of analysis**
Distribution network design	One to five years	None, or years	Maximize net revenue or return on assets	Major studies once a year; special studies if needed
Distribution network planning	One to twelve months	Days, weeks or months	Minimize total costs of meeting forecasted demand or maximize net revenue by varying product mix	Once a month
Transportation planning	One to thirty days	Minutes, hours, days	Minimize myopic distribution costs	Once a day and even driven rescheduling during the day.

Considerations regarding resource utilization, customer responsiveness, or flexibility are sometimes included in the distribution management literature next to cost-based objectives or profit-based objectives.

Not only do the long-term, mid-term, and short-term planning levels relate to diverse planning decisions and their associated planning horizon, but also, they are-

1. Strongly linked to hierarchical or graded levels in the organization; and
2. Markedly different in terms of the models that are established and executed in planning systems to support these results and decisions.

Clearly, some of these alterations have to do with how comprehensive the time aspect is exhibited, if included at all. Also, the time distribution managers spend on analyzing the solution differs significantly; tactical and strategic decision-making often comprises extensive scenario analysis based on the modeling work, while operational or operative decision-making desires speedy solutions and the likelihood to replant on an ad-hoc basis (Figure 3.5).

Figure 3.5: Various challenges in food systems.

Source: https://in.pinterest.com/pin/480829697696533754/

3.5. FOOD-SPECIFIC ISSUES IN DISTRIBUTION MANAGEMENT

In this section, we will discuss about the three food-specific aspects and their interdependence on each other. As mentioned in the introduction also these three aspects are (Table 3.2):

1. Food safety;
2. Food quality; and
3. Sustainability.

Table 3.2: Essential Requirements for Quality and Safety Assurance in the Food Industry

S. No.	Features	General requirements
1.	Quality management	Top management commitment; Plan and share responsibilities; Assign proper human and material resources; Ensure an effective quality system.
2.	Staff	Create an appropriate organizational framework; Describe the key positions; Train the workers; Develop behavioral and attitude competencies; Supervise personal hygiene and health.
3.	Production areas and equipment	Ensure proper facilities and equipment – constructions, installation; maintenance, environmental conditions, sanitizing, cleaning, validation of cleaning.
4.	Documentation	Document, develop, approve, update, distribute, and archive documentation.
5.	Processing	Validate the processes; Prevent cross-contamination during manufacture; Purchase good raw materials; Ensure the quality of the process, for intermediate, bulk or end products; Comply with legal labeling requirements; Ensure good quality of the end products (quarantine, storage, handling, delivery); Track the products rejected recovered and returned (tracking and identification).
6.	Quality control	Apply good practices in the laboratory; Apply sampling techniques; Validate the analytical method; Inspect the process; Maintain, check, and calibrate the measuring and monitoring devices.

7.	Regulations	Apply mandatory requirements; Follow contractual requirements.
8.	Consumer complaints	Handle the complaints; Document the Withdrawals / Recalls; Analyze the decisions.
9.	Self-inspection	Perform internal audit; Check compliance and corrective action.
10.	Supplier relationships	Identify and select key suppliers; Communicate clean and open; Share information and future plans; Establish joint development and improvement activities; Inspire, encourage, and recognize improvements and achievements.

3.5.1. Food Safety

Food safety normally denotes the prevention of illnesses and infections that result from the consumption of contaminated food. Because of a lot of legislation that has been enforced on this matter there is increasing attention in the industry for food safety, but it also has an economical drive: "food safety (or related information) can be a competitive factor, and more importantly, the implications of a major food safety failure can be commercially devastating." This comprises product recalls, damage to reputation as well as disciplinary liability damages (Hobbs, 2006). A familiar instance of such a food safety crisis is due to the presence of salmonella in peanut butter in the USA. The recall involved more than 200 food manufacturers downstream in the supply chain—in total recalling more than 2100 products, presence of salmonella in peanut butter was the largest product recall ever in the history of the country (Terreri, 2009) (Figure 3.6).

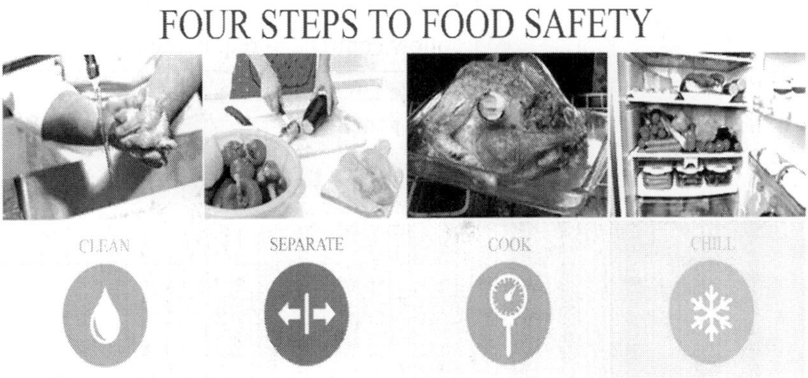

Figure 3.6: Four steps to food safety.

Source: http://speechfoodie.com/food-safety/

It has been shown that there are also a very large number of recalls of smaller scale, in addition to larger scale famous recalls (like the one mentioned above), which can also lead to serious illness or death (Salin et al., 2006). Over the past decades there are various systems and standards that have been developed in relation to food safety. The best-known are the hazard analysis critical control point (HACCP) system (FAO, 2003), the British BRC standards (British Retail Consortium, 2004) and the ISO 22000 standard (ISO, 2005). To manage food safety systems like HACCP are developed, based on risk management principles and cover a variety of physical, chemical as well as biological hazards. The basic idea behind a HACCP system is to provide a structured way to identify food safety risks and reduce or eliminate these. Standards such as British BRC standards and ISO 22000 standards typically comprise HACCP aspects, but also provide a management system to integrate food safety in an organization (Figure 3.7).

Figure 3.7: The food safety chain.

Sources: https://www.greenmelocally.com/events/food-safety-summit/

Currently, the development of HACCP system is quite common in large food manufacturers, but smaller business still have very limited usage of such systems, especially in the foodservice sector (Taylor 2008a). "It should, however, be noted that there have recently been activities to improve this situation by developing (and validating) an alternative food safety management system that is tailored to the foodservice sector" (Taylor, 2008b; Taylor and Taylor, 2008). During all stages of production, processing, and distribution, Government impose legislations that enforce traceability of food products (e.g., European Parliament and Council, 2002). Despite the importance of traceability, the truth is that in multifaceted, unified food supply chains, whole traceability is more the exception than the rule (Miller, 2009). In 2003, Schwägele argued that "traceability has to be in food companies' interest, and not just seen as legislation that has to be followed." Some recent literature follows this by discussing "how the introduction of traceability might actually be used to add value to the operations of a company" (Wang et al. 2009a). Numerous factors relevant in relation to food safety dangers relate straight to distribution management. In 2010, Van Asselt et al. found that in an extensive list of critical safety factors, for instance, the number of chain participants and the distribution of products to be of particularly strong impact.

3.5.2. Food Quality

Food quality is the second most important characteristic in food distribution chain. As noted by Grunert in 2005, "it normally refers not only to the physical properties of food products, but also to the way the product is perceived by the final consumer." For example, this can include texture or flavor in addition to the microbial aspects.

Since the quality of product in food industry is very important, Trienekens, and Zuurbier in 2008, expect that quality assurance will rule the process of production as well as distribution, and that the costs for documentation, reviewing, and quality declaration may induce responses like technological innovation to create advanced efficiency and reduce costs (Figure 3.8).

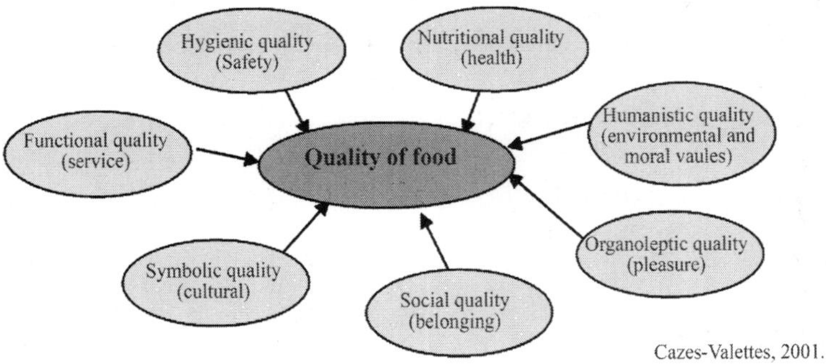

Figure 3.8: The different components of food quality.

Source: http://journals.openedition.org/aof/204

To improve temperature monitoring throughout the distribution system, new technological developments such as time-temperature integrators or indicators can be used (e.g., Giannakourou and Taoukis, 2003). This also allows for better shelf life estimation with a chain perspective, as is for example shown by Raab et al. (2008) for pork and poultry chains and Dalgaard et al. (2002) for fish chains. The culinary quality of meals in the foodservice sector, is a much-debated issue throughout Europe (e.g., Hartwell et al., 2006; Wright et al., 2006; Mikkelsen et al., 2007), and is particularly relevant considering the probable rise in consumption of professionally prepared meals (Figure 3.9).

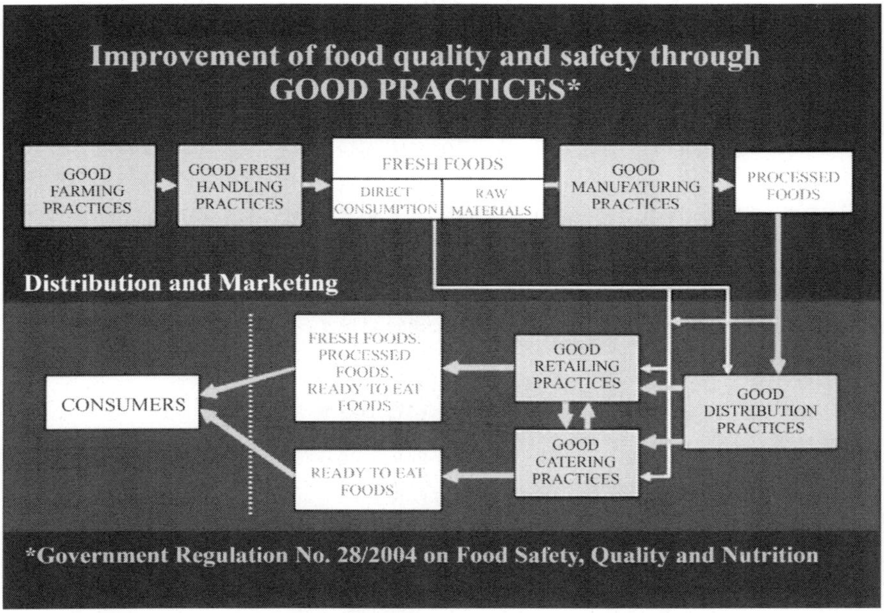

Figure 3.9: Improvement of food quality and safety through good practices.

Source: https://pt.slideshare.net/Adrienna/food-regulations-in-indonesia?next_slideshow=1

3.5.3. Sustainability

Sustainability has gained increasing importance in the food industry over the past few years (e.g., Mattson and Sonesson, 2003). "Sustainability commonly refers to how the needs of the present human generation can be met without compromising the ability of future generations to meet their needs" (WCED, 1987). It is increasingly obvious that market and controlling sustainability drivers shape or form the organization and process of supply chains. Food supply chains are at the forefront of this development (e.g., Wognum et al., 2010; Vasileiou and Morris, 2006).

Sustainability includes environmental aspects as well as a social dimension, in addition to commonly used cost-based performance measures (Kleindorfer et al., 2005). For instance, this involves, employees' health as well as safety, ethical trading in procurement of raw materials, and animal welfare (Figure 3.10).

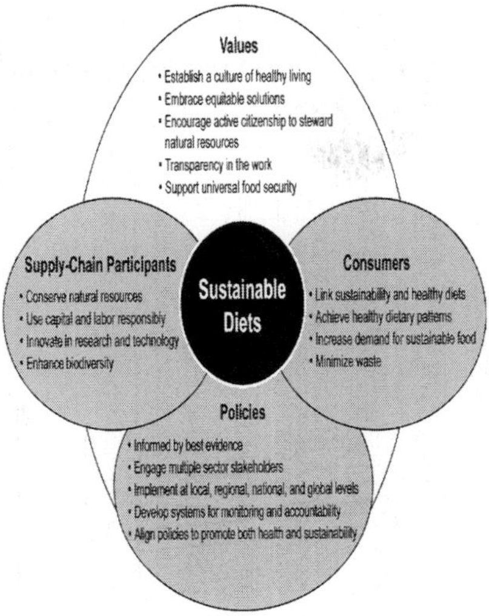

Figure 3.10: Elements needed for sustainable diets.

Source: https://health.gov/dietaryguidelines/2015-scientific-report/10-chapter–5/

For example, to improve the position of food producers in developing countries, fair trade initiatives have been developed. sourcing sustainable products from food manufacturers is also of major impact (Baldwin 2009b) following the direct influences caterers and retailers have on sustainability (among which temperature control and distribution are main parts) also related to waste and refrigeration-related to storage, and foodservice operations during preparation and service (Turenne, 2009) (Figure 3.11).

Labeling the food products is another well-known concept in relation to the sourcing and the sustainability of food chains, for instance in the form of food miles, which relate to the distance a food product has traveled to get to the consumer. Although this does not completely reflect the carbon footprint or even total environmental impact of the production and distribution system, the notion has become comparatively popular and common (Saunders et al., 2006; Wilson, 2007). It should be noted that any valuation of sustainability must essentially be made for the supply chain as a whole. When the environmental burden is high, benefits of local products in terms of food miles may get lost through a production or storage stage (Weber and Matthews, 2008). This assessment may also depend on whether a product is 'in season' or not, as this may have a huge influence on the energy usage during storage (Sim et al., 2007).

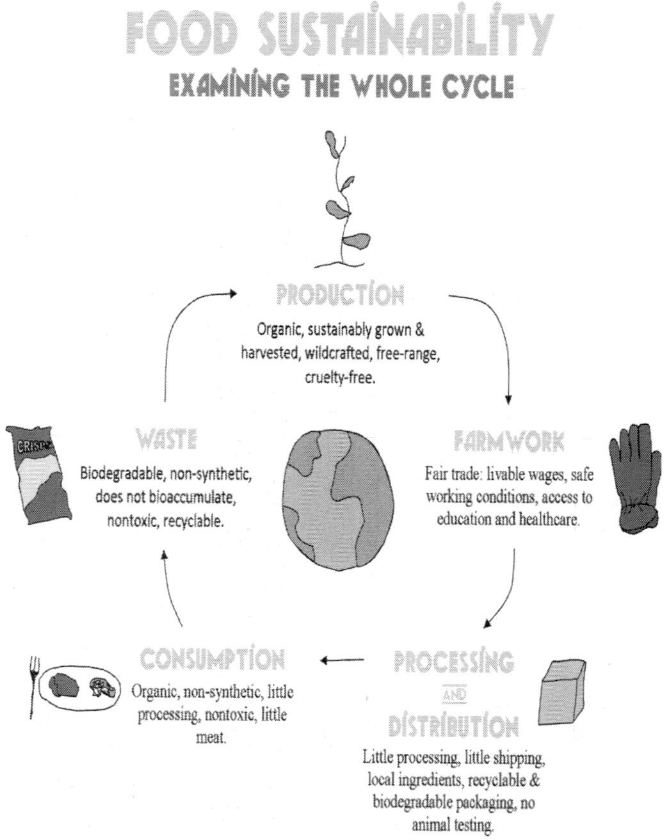

Figure 3.11: Food sustainability examines the whole cycle.

Source: https://www.pinterest.co.uk/pin/349733345774126O3/

The utmost attention has been probably received by the environmental dimension of sustainability. One of the best-known instances is "Life Cycle Assessment- an analytical tool that helps in assessing a product's environmental impact from product development to consumption" (Hauschild et al., 2005). To decrease the environmental load of products these assessments can be and are used but there is still a need of further standardization to improve relative studies and to extend practical applications in the food sector (Roy et al., 2009) (Figure 3.12).

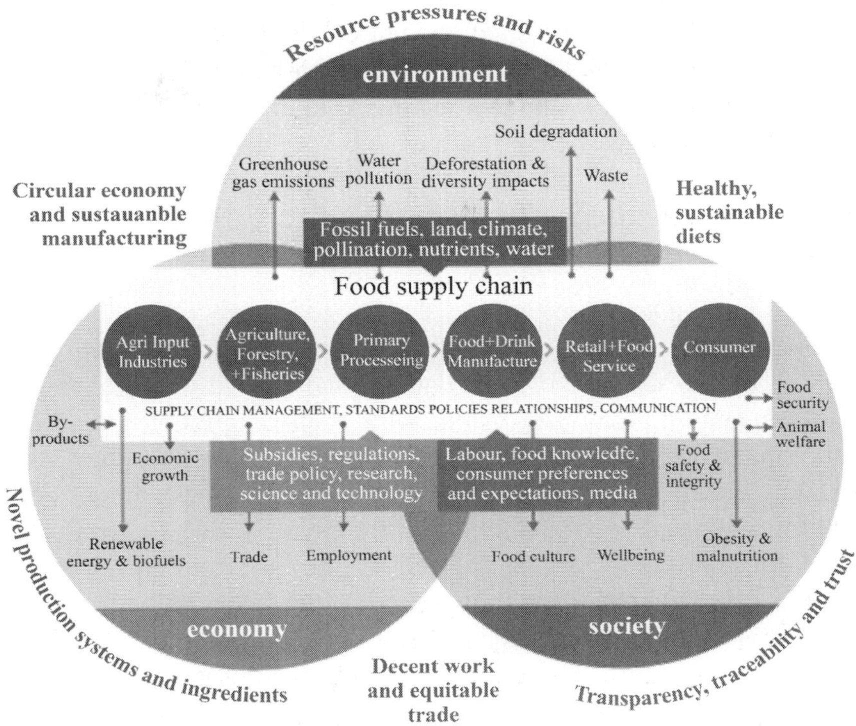

Figure 3.12: The entire food supply chain.

Source: http://www.3keel.com/sustainable-food-systems-framework-advanced/

The social dimension has received less attention in the literature (Lehtonen, 2004). Many businesses have started to work on these matters under the label "Corporate Social Responsibility" (CSR). In most of the cases, their CSR performance is communicated to stakeholders like employees or customers. For instance, as stated by Jones et al. in 2005, "all ten major retailers in the UK have stated that they see CSR as an integral element of their business environment, although there are substantial variations in the nature and extent of the CSR information they provide." However, recent work has provided some plans and procedures on how to approach this with a combination of practices to ultimately be able to combine all three dimensions (White and Lee, 2009). Also, there are several developments towards a "social LCA," which is supposed to supplement the present LCA practices and as such work towards a practice that would address all size and scope of sustainability (e.g., Hauschild et al., 2008; Hutchins and Sutherland, 2009). Promoting environmental as well as social awareness can also be valuable or advantageous for a corporate image (Chinander, 2001), and in some occasions, it might lead to price saving at the same time, for example, while reducing food waste.

3.5.4. Interdependence between Quality, Safety, and Sustainability

The above discussed three topics, that is, food safety, food quality and sustainability are presently ruling the public discussion in relation to the food sector. It is very important to understand that all the three topics are interdependent on each other, that is, there is a strong relationship between these topics. The fact that food distribution deals with products that are ultimately meant for human consumption results in a strong consideration on food safety and food quality. This is especially true for products that remain 'fresh products' with short shelf lives and do not undergo extensive shelf-life-extending treatments. Both quality and safety in the food product are based on diverse changes that occur at distinct levels. The chief change being that food quality is based on a more constant process of degradation, whereas safety is demonstrated as a binary; a product achieves protection requirementsd²E, or it does not. However, underlying is often a constant process linked to, for example, the growth of a certain bacteria temperature-controlled storage and distribution often reduce the changes that occur in products, which, however, generally need a substantial amount of energy, thereby destructively distressing the environmental influence of the products (e.g., James and James, 2010). "Twinn (2007) discusses the challenges the cold storage and distribution sector faces with respect to environmental concerns and increasing electricity costs." She stresses on the fact that there is an equal need to optimize the existing machinery and processes, in addition to the new technological solutions that have been developed. Currently, systems that are formerly designed to control food safety (such as HACCP) are also used to increase the product quality throughout the supply chain (Panozzo et al., 1999). This also concerns nutritional quality, as can, for example, be seen in the current development of the nutritional control points (NCP) notion (Rodrigues et al., 2010). This is built on the HACCP system, and can be used to recognize the serious facts in production and distribution systems linked to nutritional product variations and finally helps in increasing nutritional quality. Some of the challenges related to the sustainability of the food industry are- extending the quality and safety control systems into transparent food chains that are able to supply affordable food with high quality and diversity (Fritz and Schiefer, 2008). As stated by Smith in 2008, "the challenge for the industry is to extend responsibility for product quality into social and environmental performance of food supply chains." In 2010, Wognum et al. investigated "how existing technology designed for enhancing the transparency of food supply chains such as the safety-focused traceability systems can be expanded to also help to improve sustainability."

Additionally, with a variety of relationships the food distribution is swiftly moving towards globally inter-connected systems. This means that products are obtained from marketplaces that are located all around the world, leading to a focus on quality preservation, and at the same time cost-effectiveness. In 2005, Apaiah et al. therefore argue that the design of such food distribution systems needs a joint effort from the food process technology field, environmental science field, operations research field, marketing, and business economics field. In light of sustainability, wasted product is also an important performance measure. In general, there is a significant environmental impact on the food products that are not ending up being consumed without adding any value. Partly, this is due to food products deteriorating and having to be thrown away. As observed by Chapman in 2010, "In most countries, around 30% of food products is said to be wasted throughout the supply chain." Even though a large part of this waste occurs at the final consumer, there is a significant contribution from retail and foodservice side also. All in all, the aforementioned aspects illustrate the strong interdependence between the control of food quality and food safety on one hand and sustainability on the other hand.

3.6. DISTRIBUTION NETWORK DESIGN

Distribution network design concerns long-term decisions on the physical distribution structure of a new network or on the restructuring of a current network. For example, it comprises of the sizes number and location of warehouses and cross-docking points, as well as the associated transportation links. As defined by Jayaraman in 1998, "Distribution network design is among the most critical operations management decisions facing a firm, as it affects costs, time, and quality of customer service."

The foremost decisions are generally-

i. Where to locate facilities; and

ii. How to allocate customers to facilities and facilities to each other in case of supply chains with multiple levels. Together, this is generally referred to as facility location-allocation (Meixell and Gargeya. 2005; Melo et al., 2009).

Typically, the location-allocation problem leads to mixed-integer linear programming models in which binary decision variables are used to decide whether a possible manufacturing plant or possible distribution center is actually going to be used. Continuous decision variables are used to signify the combined product flow in the distribution network ending at the customers where request has to be satisfied. Typically, the objective function reduces the whole price for opening amenities and services at certain places and the production and distribution prices for transporting products through the distribution network.

Alternatively, profit maximization is used and to a fairly restricted amount other objective such as customer responsiveness, flexibility, resource utilization, and robustness are included, but then mostly in addition to financial facets. It should be noted that the basic single-period model best describes the location-allocation problem. The net present values of the included costs have to be used when longer time horizon and multiple periods is considered. In today's globalized situation, one might also have to extend this with, for example, exchange rate limits as well as different taxation rules and guidelines (Meixell and Gargeya, 2005).

In the food industry and distribution operations the profit margins are very low because of the significant portion of total supply chain costs, great efforts and investments are often put in network design. "It is, however, a challenging task to design food supply chains for products that have superior quality, are not expensive and are environmentally friendly" (Apaiah et al., 2005).

The eventual safety of the food product is significantly affected by the network design, the role of the network design is such that it determines the number of players, as well as the degree to which products are dispersed through the network. Regarding product quality, the design of the network influences, for example the time a product is subject to quality degradation during distribution. In terms of sustainability, there is a very strong impact of the design of distribution networks, for instance related to the environmental impact of the transportation method involved or to the distance products have to travel to reach the final consumer. For foodservice chains, the aim is often to a no-stock overnight policy in distribution centers or to pursue a low stock level, and to regularly ship in lesser quantities with high disparities in demand. This ultimately will affect the length of the chain: for instance, direct delivery from the producer to the caterer is common for these products (Bourlakis and Weightman, 2004). Production activities are not always confined to the initial food manufacturing stage, which is another typical aspect found in foodservice systems, which is often the case that additional production steps take place at the caterer (e.g., final meal assembly and preparation).

3.7. PACKAGING AND LABELING

Both packaging and labeling are included in a quality control program. Packaging and labeling are the first items that influence the consumer to buy any commodity. If the packaging and labeling of the product are attractive in appearance it automatically captures' the customers' attention. At the same time, if packaging or labeling is poor or not so good it may create negative impressions relative to product quality.

There are two basic packages that are typically needed for food products:

- The primary package encloses the food. This package has direct contact with the product. A bottle, box, carton, film, or jar are some of the common primary packages; and
- The secondary package is used to collect numerous packaged food items for delivery. The objective of the secondary or shipper package is to provide protection, reduce handling of each individual carton or bottle and is needed for efficient movement of goods to the consumer.

There are usually three components in microwaveable products and other packaged foods, these components are: the carton, the pouch, and the shipping case. This is true for both simple and complex packages or labels. The basic purpose of packaging is to protect the food product as well as to allow the product to get shipped to multiple units without damaging the product inside the package.

Functions of packaging:

1. Promoting and selling the product;
2. Defining product identity;
3. Providing information;
4. Expressing customer needs;

5. Ensure safe use;
6. Protecting the product.

Items for packaging consideration are:

1. A statement from the supplier that the packaging is made of USDA and/or FDA permitted materials. The statement should list all the package composition.
2. Dimensions of bottle, box, carton, or jar.
3. Strength of the container and suitability for freezing, microwaving, or stacking.
4. Strength of seals or fit of the lid. The temperature requirements for sealing are critical for heat-sealed packages.
5. Ability to restrict or allow air flow, moisture or light. Flexibility, permeability, thickness, and temperature resistance are exact principles in this category.
6. Graphics (illustration, picture or visual designs).
7. Label format as well as legal requirements.

Based upon the particular food item, packaging must be selected or designed. The packaging of fresh fruits and vegetables requires protection while allowing air flow for proper refrigeration and respiration. The packaging of dairy products requires to inhibit light penetration and excessive oxygen because of the potential for flavor flaws or defects due to oxidation, rancidity, or the absorption of foreign flavor. A final instance, the tea bag must provide permeability to moisture.

Package graphics, by pictures or words, outline the contents and assist as a point of purchase information. As per the government laws, the package requires all information like- product name, ingredient statement, and manufacturing or distribution location. Governments' rules and regulations list numerous requirements for packaging and even extend to specifying the size or type or printing. Pictures, graphics, or any other thing printed on the package are optional and serve to inform the consumer. To avoid mislabeling or misbranding, graphic or picture designing is done to represent the contents of the container.

Some typical packaging and labeling defects include broken seals leaks, color variations, scuffs, smears, short fill and product infestation or spoilage. The possibility of defects can vary from single unit packs to multi-packs (shippers). It is to a food processor's advantage to develop packaging and label specifications along with the same format as ingredient specifications.

When there is involvement of direct contact with the food, materials of construction are chiefly important. The food products may get contaminated by certain chemicals or foreign materials from packaging materials (Figure 3.13).

Figure 3.13: Food packaging.

Source: http://podshifter.com/food-packaging-of-the-future/

The packaging material must meet USDA and/or FDA requirements. Use a reputable or trustworthy packaging supplier. The manufacturer of the package is the chief source for package specifications. Local distributors can get the desired information from the manufacturer. To prevent problems such as under or over-fill; spillage or breakage of the container; shifting within the package, dimensions of the package, both inner and outer, are defined. If the control process is absent there are very high chances of product loss, giveaway, or lost sales. The strength of the container and the seals or the fit of the lid are significant considerations. If there is any negligence with regard to these items it can result in breakage, crushing, or spilling. Most important is the potential for microbiological or physical contamination when a poor seal of the improperly fitting cap is a packaging defect.

These are the items on the label that are required by law.

- manufacturer's name and contact details;
- the name of the product;
- description of the product;

- weight (some foods are exempt, for example, bread);
- ingredients (listed in descending order of weight);
- cooking/heating instructions;
- storage instructions;
- shelf life;
- place of origin;
- allergy information.

The following items are not legal requirements, but are nevertheless good practice and often included on packaging (Figure 3.14):

- illustration of the product;
- price;
- the nutritional values of the product;
- customer guarantee;
- the batch-code and bar-code numbers;
- opening instructions.

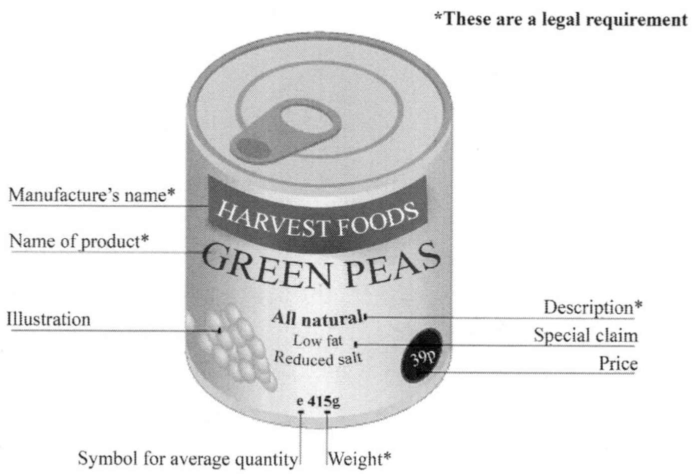

Figure 3.14: Food labeling.

Source: http://www.bbc.co.uk/schools/gcsebitesize/design/foodtech/packaginglabellingrev5.shtml

REFERENCES

1. Akkerman, R., Farahani, P., & Grunow, M., (2010). Quality, safety, and sustainability in food distribution: a review of quantitative operations management approaches and challenges. *OR Spectrum*, *32*(4), pp. 863–904.
2. Bianchi, M., (2012). *Balancing Food Safety and Sustainability Opportunities for Co-Management.* [eBook] Available at: http://ucfoodsafety.ucdavis.edu/files/150364.pdf [Accessed 17 Mar. 2018].
3. Daani, S.(2015) *Food Supply Chain Management and Logistics.* [eBook] Available at: https://www.koganpage.com/media/project_kp/document/food-supply-chain-management-and-logistics-sample-chapter.pdf [Accessed 17 Mar. 2018].
4. Food Processing and Quality Control, (2007). [eBook] Available at: http://documents.wfp.org/stellent/groups/procumeeting_content/documents/events/wfp126925.pdf [Accessed 17 Mar. 2018].
5. Keeping Food Safe, Sustainable, and Socially Responsible, (2013). [eBook] Available at: https://www.bsigroup.com/LocalFiles/EN-AU/Food%20Documents/BSI%20ANZ%20Food%20Portfolio%20Brochure_online.pdf [Accessed 17 Mar. 2018].
6. Knaflewska, J., & Pospiech, E., (2007). *Quality Assurance Systems in Food Industry and Health Security of Food.* [eBook] Available at: https://www.food.actapol.net/pub/7_2_2007.pdf [Accessed 17 Mar. 2018].
7. Nordenskjöld, J., (2012). *Implementation of a Quality Management System in Food Production*. [eBook] Available at: https://stud.epsilon.slu.se/4676/7/nordenskjold_j_120815.pdf [Accessed 17 Mar. 2018].
8. *Quality Control a Model Program for the Food Industry*, (2010). [eBook] Available at: https://athenaeum.libs.uga.edu/bitstream/handle/10724/12251/B997.pdf?sequence=1 [Accessed 17 Mar. 2018].
9. Rajendran, M. (2011) *Concepts of Food Safety and Quality Management Systems*. [eBook] Available at: http://old.fssai.gov.in/Portals/0/Training_Manual/Presentation%20on%20concepts%20of%20Food%20Safety%20and%20Quality%20Management%20Systems%20by%20Mrs.%20Malini%20Rajendran.pdf [Accessed 17 Mar. 2018].

10. Yamina, F., Mebarek, D., & Noureddine, B., (2014). *Food Risk Management and Sustainable Development*. [eBook] Available at: https://file.scirp.org/pdf/JSSM_2014043010065191.pdf [Accessed 17 Mar. 2018].

CHAPTER 4

SUSTAINABILITY IN SUPPLY CHAIN AND FOOD INDUSTRY

"Almost all quality improvement comes via simplification of design, manufacturing, layout, processes, and procedures."

—Tom Peters

CONTENTS

This chapter speaks about sustainability and the steps to reach over there for the food industry. It also answers the question like why sustainability in supply chain and food industry is required and how to attain this with focusing on manufacturers, farmers, retailers, and consumers the end user of the produced materials. In order to achieve sustainability related to supply chain and food industry, each one in the supply chain has to join hands together and to strictly follow the set standards by the regulators. The ways discussed in this chapter are an increment in production by using lower resources and providing the highest value to the consumer so that they could be turned into long-term consumers. The expected learning out of this chapter is to acknowledge the requirements of sustainability in the food industry and ways to save the environment for our younger generations.

4.1. INTRODUCTION

Today businesses and industry are enlarging the boundaries of their business performance beyond financial parameters and contributing towards sustainability. Corporate stakeholders increasingly demand businesses to consider and account for sustainability implications of operations beyond the organization and across a supply chain context especially in the food industry.

Although there are remarkable improvements in the field of sustainability benchmarking in the food industry to account for lifecycle influences of products, services, and processes, there is still considerable room for tools and methodologies to effectively measure and benchmark 360-degrees sustainability performance of supply chain operations in the food industry.

One of the major challenges for sustainability benchmarking across supply chains is to account for the multi-dimensional nature of sustainability encompassing profit, people, and planet commonly known as 'triple bottom line'/3-dimensional model implications of business and supply chain operations.

This chapter reflects three dimensions of sustainability – environmental, social, and economic following a concept of triple- bottom line and applying sustainability indicators to various supply chain stages from production to retail in the food industry by incorporating strategic sustainability target setting through scaling indicators according to the desirability of sustainability performance. The food supply chain serves as a useful example because sustainability implications of food production and consumption are critical for many food organizations, , and other stakeholders, who are highly concerned with sustainability effects of supply chain operations in the food industry (Figure 4.1).

Figure 4.1: Sustainability in food industry.

Sources: https://assets.tvm.com.mt/mt/supermarket

Every year, the food supply chain emerges as more sustainable, based on the various reports that measure the success of such initiatives. Food manufacturers, distributors, and retailers boast on their outstanding achievements in annual sustainability reports of their business. While all the players are sharpening their efforts, the companies that facilitate product movement from manufacturer to consumer that is logistics service providers play a crucial support role, particularly in the area of visibility/trackability. But are these initiatives sufficient enough in tackling the challenges of growing population, climate change and declining agricultural production? Is there a uniform measure by which governments and organizations can monitor progress in improving sustainability for their businesses? What is the supply chain's role related to food industry in addressing these challenges? Many of the sustainability challenges such as water usage, deforestation, genetically modified organisms, worker's treatment, declining farm space, increasing investment in agricultural fall more on agriculture and processing than on the movement of food. Few observers argue, however, that the supply chain plays an increasingly important role as companies recognize the importance of supply chain visibility in improving their sustainability efforts along with other processes.

According to Pan Chen, vice president of analytics and supply chain services at HAVI Global Solutions, the supply chain consultancy "Sourcing raw materials is a huge part of the supply chain function, while the actions may take place at farms and in manufacturing facilities, it is up to supply chain professionals to determine where and how much to source raw materials." For this, HAVI explores opportunities to reduce waste through more dynamic and sustainable supply chain optimization, Chen says. "This includes determining the most favorable geographic layout of supplier locations, manufacturing facilities, forward warehouses and distribution centers to maximize efficiencies. And when market conditions change and there are fluctuations in demand and availability of supply, we adapt the supply chain accordingly and make adjustments in real time. Getting the underlying infrastructure right to provide the type of transparency that consumers need should be a priority."

According to Skip Miller, vice president of customer service and solutions at CHEP USA, global provider of container and pallet pooling services, told a group of consumer food industry executives earlier this year that the optimum way to identify small changes that can boost savings and growth is for trading partners – manufacturers, retailers, growers, and suppliers – to evaluate their supply chain to identify gaps, overlaps, and inefficiencies. Once those inadequacies are identified through supply chain mapping, experts can develop solutions that can help save money, be more efficient and sustainable.

CHEP USA, for its part, has cooperated with consumer supply chain experts to identify inefficiencies in its supply chain and to develop cost-effective solutions for the industry. These efforts include the following use cases:

- A wholesale business improved dock space and transportation efficiency by diminishing multiple platforms and vendors.
- A beverage manufacturer saved $500,000 annually by changing its primary and secondary packaging for the products.

- A Canadian retailer saved $75,000 annually by streamlining the handling and transportation of fresh produce from the U.S., Mexico, and South America.

4.2. NECESSITY OF SUSTAINABLE SUPPLY CHAIN IN FOOD INDUSTRY

The population in the world is growing rapidly and is expected to surpass 9 billion by 2050. That's an extra 2 billion population that must be fed and clothed every day amid an increasing scarcity and conflict over land, water, and energy resources which are essential to live a life on this earth. "In order to provide adequate, safe, and nutritious food to all, production must increase by at least 60% per year. People working in food & agriculture industry, the challenges have never been resolved as they appear again and again. These challenges include the impacts of climate-related disasters such as droughts in the African region and Asian as well as increasing energy costs, increasing waste and eco-system damage across the globe. Traditional and inefficient farming practices in the world, including low-technology and high manpower-based farming, complicate the picture," according to the *Food and Agriculture Organization (FAO)*. As per the research conducted by FAO, agriculture is expected to produce more output than existing while reducing its impact on the planet's natural and built environment. To make it possible countries must be going by promoting sustainability throughout the value chains (from manufacturer to consumer), and focusing on environmental, social, and economic issues, the 3 factors of sustainability.

While the problems may seem intractable among industries, many innovative businesses like Ikea, Amazon, etc. are already delivering win-win outcomes in terms of financial returns (profits) and sustainable development. These breakthroughs innovation of technology and sustainable ideas not only maximize productivity for the businesses, but also improvise delivery across a far more complex landscape of sustainability factors – financial, environmental, and social serve. In order to optimize processes that provide our daily food including farmers, traders, processors, and supermarkets and the scientific and political communities and interest groups cooperation between every parties in the supply chain is essential and it will create more value to the industry. The private players in the industry require innovative expertise, networks, appropriate sustainable solutions and valuable business partners collaboration, including financial services providers for the long term.

4.3. IDEAS TO DEVELOP SUSTAINABLE SUPPLY CHAIN FOR FOOD INDUSTRY

Food & Agribusiness (F&A) companies are facing increasing structural process related pressure on their supply chains that can be visualize globally, as a result of that price volatility increasing, margin pressure shifting along the chain, and an increased complexity in their operating environment which include 'Planet, People, and Profit.' Complexity includes even external agendas such as sustainability for the industry, health/safety issues of related people and the increased stress on resources from food, feedstock, and biofuels. The current view of supply chains in the food industry is of a zero-sum game that leads to requirement of innovation. Especially, Downstream companies (such as large food companies and retailers), who buy agri-commodities for further processing, are under huge pressure to protect margin in a more complex and competitive environment, fluctuating global demand and volatile prices for the end products. Their major issue is how much of the margin could be shared to secure supply of the products. On the other hand, upstream suppliers (small food companies and retailers) are strengthening their hold on competition and negotiating position. They look for higher prices for the products to cover increasing production costs (Variable cost) and are demanding more from a buyer in return for an assured supply commitment which makes them satisfied and long-term consumer. As a solution to the complexity, companies in the industry should start looking for ways to maximize their control over the supply chain and align their chain with strategic growth priorities. In order to deal with this complexity and pressure a common topic of discussion with all senior management of F&A companies is required. Supply chain management has become a strategic issue for the industry as well as organizations – companies are engaging more deeply with suppliers and their chains to resolve this. Smallholders, retailers, sellers are a particular focus.

The strategy to win this is cooperation between stakeholders related to the industry in the supply chain leading to the circular beneficial economy model for the long term. The complete chain should work in collaboration to maximize the value of outputs and turning waste into new cash flow opportunities like recycle management. In an empowered and dedicated supply chain, F&A players in the industry make a collaborated and formalized long-term commitment to working together on the complexity, which generally extends end-to-end up to the final stage that is customer, along the supply chain. It is aimed by the companies in the industry to build new levels of trust and shared ownership for innovation in the processes and products. The explaining characteristic is the commitment to work together to satisfy requirements of a 'captain,' which typically is a large downstream company (larger one) – either a processor/brand owner, retailer/food services operator or exporter in the industry. To attain this, it should be kept in mind that the level of integration and cooperation among companies will vary from one commodity chain and geography to others, depending on several factors, including how tight/improve the chain is. We can discuss an example of sugar or coffee, typically have tighter integration between producers and processors, vis-a-vis cereals. As key member of the chain, the major processor, retailer, foodservice player or exporter acts as a captain to showcase leadership and providing commitment to its suppliers.

There are some proven benefits for the collaborative approach, including:

1. Reduced supply risks of parties;
2. Improved productivity and profitability;
3. Access to new markets and capital (market expansion);
4. Enhanced brand reputation for betting on new population for the company.

4.3.1. Ways to Improve Sustainability in Food Supply Chains

- **Increment in Output with less land and few resources**

  Agricultural players in the food industry and food supply chains will need to increase their production output, but with consuming fewer natural resources and with respect for both human and animal welfare for attaining greater sustainability. Companies can do this by providing access to finance dedicated to sustainability and to its expertise and networks.

- **Increased Value Addition for the Consumers**

  Utilize the valuable insights and knowledge gained from the global initiatives to add real value for the customer the real asset for the company – who might be anyone from a wheat producer in California to a supermarket chain owner in Brazil to a dairy farmer organizer in New Zealand to a Dutch potato processing company. Customers are the end user of your product and if they are satisfied with your product you may grow like anything. Sustainability can drive all the product/services we provide to our customers in the food industry and across the food supply chain as a whole, 'from seed to eat.' Companies should provide its customers with the following services:

Access to finance: Production requires capital more or less. Companies should have set a number of sustainability criteria that customers must comply with in order to be eligible to receive financing support for the farming. These kinds of business operations have consistently proved to be successful in capitalizing on opportunities. They turn into stronger companies with improved risk profiles. With its focus on the long-term, companies must grant access to finance to small and big users.

Access to knowledge: Companies must grant the access to knowledge to their customers worldwide operating in the agricultural and food production industries. It could be technical expertise or how supply chains and markets operate in food industry. They must invest in knowledge of, and research into, the economic aspects of food production and how to improve its sustainability. Investment in improving people's skills and raising their awareness may turn customers into long-term customers. Food industry players in the industry must share the knowledge with customers through formal or casual presentations, customer meetings, publications, and various classes and seminars.

Access to networks: Establish links between chains for various products and participants of the supply chain. Ties to international policymakers, universities, and research institutions to assist customers and accelerate change can be a possible way to have a good access to networks.

- **Establishment of Regulator on the Global Stage**

  Companies in the industry should be active participant worldwide in various round tables to improve the sustainability of agricultural production chains. The members of the Round Tables could be – producers, customers, social groups and scientists – and to meet regularly to discuss the set standards for the sustainable production of commodities such as crop, meat, palm oil and seafood. Regulator must also assign norms to be performed with set restrictions in the industry in order to achieve the sustainability. Companies must have co-ordination to reach to the common goal of sustainability in the supply chain and the food industry.

4.4. BUILDING TRANSPARENCY IN FOOD SUPPLY CHAIN

Farmers worldwide are in a tight spot today. On one hand of the activity, the farm needs to produce profit-making crops and run efficiently and smoothly. On the other hand, consumers are asking for transparency and sustainability in farming so that future could be retained. Digital process in transformation and innovation may seem like something that happens in other industries like IT, but industry giants such as Cargill and John Deere are proving digital business transformation can lead to success in agriculture as well. As our population grows in the future, farms that embrace precision agriculture products can see excellent gains. What kind of gains of course, in terms of sustainability and resource management. The research has found that early adopters of the industry are seeing an average 9% increase in revenue from the production, 26% increase in profitability, and a 12% increase in market value altogether.

As communication in the supply chain improves, more impact in the industry from consumer opinion from food recalls, demand for locally sourced foods, and increased supply chain transparency. From ice cream to salad dressing in the industry, recalls are causing serious concerns for consumers and problems for agricultural production due to depletion of the resources. Consumers today want farm-to-fork transparency to ensure their food safety in order to earn that much. Consumers are also more aware and concerned about agricultural practices in the industry and how their food was produced keeping sustainability in the mind. Many consumers are fine to pay more for food products that is proven to be sourced using fair trade practices, which require the farmer to follow sustainable farming practices to save future and natural resources. But the question, how do you build food traceability to that level? Smallholder (Small players in the industry) farming has answered this question in one fashion with the increasing trend of direct marketing by farmers to the end consumer with no connections in between. The farmer today can create a deep, one-to-one relationship with the consumer, who they see at farmers' markets all over the world, on CSA days, and in the local community. This increasing trend has been driven by society's sustainability demand for further transparency in the food supply chain. But clearly, this approach is not capable enough to the answer for global agricultural supply chains in the industry. Consumers in the world cannot maintain one-to-one relationships with all the farmers that produce input for their food, which provides a level of transparency while automating most of the communication needed. Hyperconnectivity to all participants helps provide related information, captured by digital farming solutions and processing practices, from the farm to the end consumer the user of the products. It suggests consumers the related information they need, from the seeds and inputs used to the processes performed. This can also ease out the effort of maintaining so many one-to-one relationships in the industry. This task may also affect the commodity markets because traders are as limited as mixing and blending products impacts traceability for the industry, side by side with transparency is sustainability. The world's population is growing rapidly and is expected to approach 10 billion people in the world by 2050, which will require a 70% increase in food production to meet their requirements. There is no surprise that scarce resources in the earth such as water and arable land are becoming more valuable in order to their depletion. Additional transparency in the supply chain and networking also provides opportunities to optimize and increase the efficiency of food supply by reducing waste. It also exceeds government, private, and NGO scrutiny, affecting increasing and constantly changing regulations on farm operations, processing, and traceability. Process transformation through digital innovation minimizes the investment cost of transparency in the food supply chain: As more businesses are leading to digitalize, they can provide more information at a lower cost to the end consumer through mobile/digital apps and QR codes. This level of digitalized transparency helps protect agribusiness and improves efficiency in farm operations (Figure 4.2).

Sustainable farming methods in the industry provide opportunities to improve profits while also preserving the environment for the future. Precision farming, the innovative technique of farming is one area where technology and sustainability collides, because inputs are used only where they are needed, reducing water, fertilizer, and fuel use. For example: in *California's* ongoing drought, precision agriculture is expected to reduce water usage on farms by 25% as usually we are consuming. Getting fresh water supply today is significant concern for sustainable agriculture. 80% of the world's arable land is watered exclusively by rainfall and produces 60% of the world's plant-based food, today. The remaining 20% is irrigated and produces 40% of the world's plant-based food. As aquifers in the world sink to record lows, the use of irrigation falls into a bad light. Precision agriculture can bring innovative technologies to better match appropriate crops to soil and climate conditions and help protect our limited fresh water supply. Such opportunities for the industry provide exciting opportunities for creativity and digital entrepreneurship in farming.

Figure 4.2: Sustainable farming.

Source: https://Sustainable farming upload.wikimedia.org/wikipedia/commons

Another area in the industry where sustainability in the supply chain in the food industry can play a role is in helping maintain and preserve the family farm. Family farms in the world are becoming rarer as younger generations, the innovative one realizes the challenges of profitability and leave in search of better opportunities. Precision farming which is based on low water usage limits inputs, automates tasks, and helps slow or eliminate the loss of experienced operators. This in turn boosts profitability in the industry, making the business of farming more appealing to younger generations the future users. Digitalizing agriculture for the sustainability improves transparency and sustainability. 90% of executives recognize the impact the digital economy has on their business today. Only 15% are still creating a plan of action to adapt to these changes.

4.5. SUPPLY CHAIN MANAGEMENT

Supply Chain Management or SCM includes the management and planning of various activities that are a part of conversion, procurement, sourcing, and all the activities related to logistics management. (Council of Supply Chain Management Professionals, 2012) that provides information and statistics on economic benefits regarding the companies. Although these supply chains are feasible in economic terms, they are not considered sustainable in long-term plans. Reason for this is the fact that they can put a negative impact in terms of social aspects and environment. To overcome this drawback, the concept of Sustainable Supply Chain Management (SSCM) has been developed in recent years. SSCM has placed itself as a transparent and strategic integration, and achievement of any company's environmental, social, and economic goals along with the systemic coordination. The coordination enables inter-organizational business processes in order to improve the economic performance in long-term, be it an individual company or the complete supply chains (Mentzer, DeWitt & Keebler, 2001; Lambert, 2006; Carter & Rogers, 2008).

There are necessary methods and tools that should be implemented in order to measure, evaluate, and compare most significant environmental impacts in the supply chains along with the improvements and progress achieved by the implementation of these initiatives. There are multiple ways to understand and study the environmental (Finnveden & Moberg, 2005) and sustainability aspects of consumption and production (Ness, Urbel-Piirsalu, Anderberg & Olsson, 2007). The basic approaches include qualitative and quantitative; further qualitative approaches cover a cleaner production, design for environment, industrial ecology while quantitative models have strategic environmental assessment, material flows accounting and Life Cycle Assessment (LCA) (Heijungs, Huppes & Guinée, 2010). LCA is the most established and well-developed tool in all of the mentioned ones (Ness et al., 2007). Due to increasing concern of people about carbon emissions and global climate change as a casual factor (Rotz, Montes & Chianese, 2010), the term Carbon footprint was coined and become popular over the last few years. Carbon footprint (CF) in definition can be stated as a procedure that is referred to the quantity of total green-house gases like methane, carbon di oxide, halocarbons, nitrous oxide, etc which is produced during product's life cycle (Pihkola, Nors, Kujanpää, Helin, Karimiemi, Pajula et al., 2010). In fact, carbon footprint analysis can be taken as a simplified version of LCA Analysis in which the major focus is only on the environmental impact of Global warming, instead of taking all environmental impacts into consideration.

From the consumer's point of view, carbon footprint is taken as a subset of growing demands for environmental information that is a key ingredient for decision making that is highly concentrated on the knowledge of subject, in terms of sustainable production and consumption (Carballo-Peneda & Domenech, 2010).

The different stakeholders can be involved regulators, decision-makers, potential investors, and these stakeholders play a strategic and crucial role as a decision-making tool for evaluation and assessment of environmental impact that FSCs are imposing on surroundings and nature. Along with understanding the environmental impact, these tools are used for investigating the actual impact of sustainable practices toward decarbonizing food networks.

4.6. SUSTAINABILITY PROGRESS IN FOOD INDUSTRY

Oxfam International, a global workers rights' organization that examines food and agriculture companies' actions, has criticized many companies' practices. But in doing so, it has also provided a yardstick, however unwelcome, by which to measure sustainability progress. The Oxfam *'Behind the Brands'* campaign analyzes and compares the various sustainability commitments of the Big 10 food companies. The campaign addresses seven specific policy issues–land, women, farmers, workers, climate, transparency, and water. Unilever recently grabbed the No. 1 Oxfam sustainability ranking from Nestlé, with Coca-Cola Co. getting the third spot. Kellogg Company has made the stupendous progress across all sustainability areas–up 30% from its rating three years ago. *"The company (Unilever) has strong policies on deforestation and palm oil, guidelines for its suppliers, and is engaging governments to take action,"* the report reads. *"We hope they will continue to show leadership and go even further."*9 of the Big 10 industry giants have improved their rating by at least 10%age points since the *Behind the*

Brand's campaign's launch 3 years ago. Danone–the only company not to surpass the 10-point barrier–still increased its score by seven points, thanks to commitments on climate (Figure 4.3).

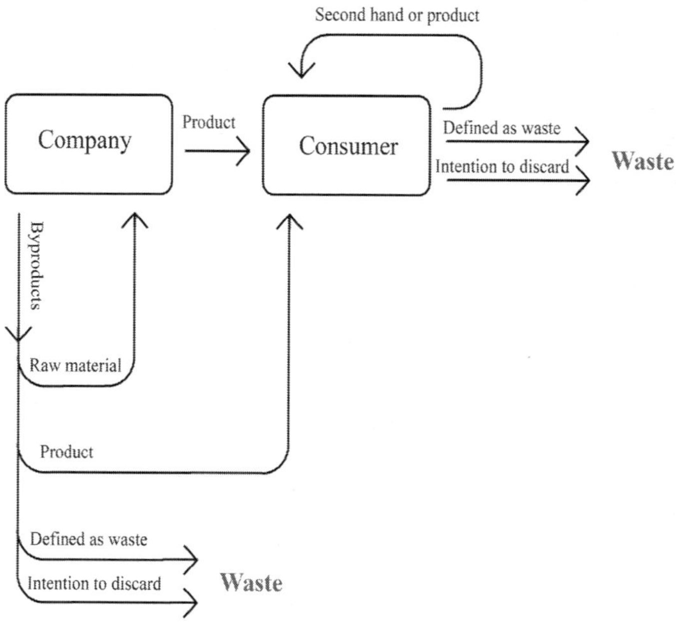

Figure 4.3: Waste Management and sustainability in food industry.

Sources: https://Sustainability/wikimedia.org/European

Shift in Market Demands

Market demands for more organic, fresh, and unprocessed food places further pressure on sustainable food and agriculture. Hebert Research Inc. found that more than 29% of consumers are opting and consuming organic, fresh, and unprocessed foods. The trade-offs of more organic food production and increased demand focuses the concern of sustainability. It is not only meeting organic standards of soil as decided by authority, but the supply of ground and surface water.

Recycling and Reuse Growth

Waste management and recycling are areas that the food industry continues to invest in. This year, GMA has conjoined with the Food Marketing

Institute (FMI) to standardize food waste practices in the industry and communicate them to consumers. The Food Waste Reduction Alliance (FWRA), a cross-sector industry initiative led by the GMA, the FMI and the National Restaurant Association (NRA), recently released an updated guide with practical steps and examples to help manufacturers, retailers, and restaurants cut food waste. The Campbell Soup Company's partnership with the Food Bank of South Jersey and the New Jersey agricultural community gave undersized peaches headed for the landfill new life as shelf-stable peach salsa. Proceeds from the sales of the salsa, "Just Peachy," benefits the Food Bank of South Jersey (Figure 4.4).

Figure 4.4: Recycling, reduce, reuse (3R model).

Source: https://cdn.pixabay.com/photo/environment

ConAgra Foods Inc. has changed the way it transitions from one pudding flavor to another in a manufacturing facility. The company now creates blended flavors rather than waste product while flushing the manufacturing line from one flavor to another. The company now donates the mixed-flavor pudding to various hunger organizations, reducing manufacturing loss and getting food to those in need. Big foodservice operators such as Aramark Corporation and Sodexo have reduced food waste by eliminating trays in cafeterias. Darden Restaurants and Yum! Brands have collaborated with the

Food Donation Connection to coordinate food donations to food banks and other organizations as an alternative to discarding prepared foods.

Sustainability Continues on Different Fronts in the Industry

Walmart recently released its 9th annual Global Responsibility Report for sustainability, outlining its environmental and social activities for the last year. For the first time, this report includes information about the progress made on implementing the company's sustainable chemistry policy. According to Wal-Mart, it has cut down the usage by weight of its designated high priority chemicals by 95%. Information about how this progress was made by the company will be posted on Wal-Mart's Sustainability Hub, along with chemicals the company has targeted for reduction or elimination. Wal-Mart's 10-year collaboration with the *'Environmental Defense Fund'* identifies three sustainability goals (Figure 4.5):

1. Be supplied by 100% renewable energy;
2. Create zero waste; and
3. Sell products that sustain the world's resources and environment.

Figure 4.5: Walmart US.

Sources: https://c1.staticflickr.com/walmart

Roundy's Inc., joined the U.S. Environmental Protection Agency (EPA) GreenChill Partnership program, which is a partnership between the EPA and food retailers to reduce emissions and decrease their harmful impact on the ozone layer and climate change. In 2014, Roundy's top the list of the grocery store in Wisconsin, and only the 3rd store in the United States, to utilize a Transcritical carbon dioxide (CO2) refrigeration system in their Menomonee Falls, Wisconsin Pick 'n Save location. The Transcritical CO_2 system eliminates the use of ozone-depleting hydrofluorocarbons (HFCs). 1 pound of leaked HFC refrigerant has more global warming potential than 4,000 pounds of carbon dioxide. The "Menomonee Falls" store is announced as one of the "greenest" supermarkets in the nation and was awarded a 'Platinum level certification' by the EPA's GreenChill Partnership program. Using the EPA's Greenhouse Gas Equivalencies Calculator, the yearly reduction of greenhouse gases are approximately 33.6 metric tons and 36,047 pounds of coal being burned. Roundy's has pledged to install the Transcritical CO2 refrigeration system or whatever improved non-HFC technology becomes available in new stores and stores where the refrigeration system is scheduled for replacement.

Fetzer Vineyards became the first winery in California to process on 100% clean energy in 1999. By 2006, the company covered 75,000 square feet of rooftop with solar panels, creating 80% of the energy needed to run its bottling facility. In 2012, Fetzer Vineyards adopted a new cleaning product called peracetic acid, which requires less rinsing as compared to the previous one, for the wine tanks. This change enabled the company to reduce its water consumption by more than 200,000 gallons annually, and reduce the energy used to pump the water (Figure 4.6).

Figure 4.6: Fetzer vineyards sustainability.

Source: https://lnt.org/sites/default/files/fetzer

Sustainability has definitely become a driving force in meeting consumer demands, government set regulations and more efficient business practices for the food industry. While all players having presence in the food supply chain have addressed sustainability, logistics service providers play a critical support role in many of these efforts.

4.7. MATERIAL HANDLING INSTITUTE OUTLINES SUSTAINABILITY TOOLS FOR FOOD INDUSTRY

By 2025, the material handling and logistics industry should have standard methods of incorporating sustainable development into business plans and operating strategies, according to the Material Handling Institute (MHI) "U.S. Roadmap." Such methods should go through the goals of sustainability, while maintaining and advancing the commercial interests of the industry. Any attempt to eliminate the environmental impact of the material handling and logistics industry requires an accurate assessment of environmental impact itself, the Roadmap notes. To this end, there is a need:

1. To specify in some unit of measure the relevant emissions released; and
2. To identify an environmental impact factor that might be expressed in monetary terms.

Consideration of environmental harmful impact could include effects such as deterioration of human health who are also consumer of that industry, loss of wildlife, agricultural effects and global warming. Any assessment of impact requires substantial care because impact depends on the context with respect to the surrounding environment society and culture. Life cycle assessment (LCA) tools already exist to help companies assess the impact of their operations on the environment. These tools should address the details of material handling and logistics for the food industry. The more challenge, MHI notes, is to spread the word within the industry that such tools already exist. One of the factors that contribute to emissions generated by the material handling and logistics industry is the long distances traveled by trucks and other vehicles for supplying goods from places to places. As the supply chain industry continues to reduce the cost of transportation by reducing miles driven, the environment will benefit and will lead to sustainability for the food industry. A natural tension even the major one, exists between shippers, who seek to consolidate shipments for transportation efficiencies (and perhaps unintentionally for environmental benefit), and consumers who want what they want, when they want it and where they want it. Through increased awareness or carefully designed pricing strategies for the industry, changes in consumer behavior could lead to reduced impact on the environment.

A noticeable amount of solid waste generated by the food industry comes in the form of packaging which contains products to the large extent. There is a requirement for new approaches to reduce the amount of product's packaging required and for new packaging materials and approaches that support future material (Plastic less) handling and logistics requirements. Packaging should be capable enough for preventing product degradation due to environmental factors, preventing physical damage of products, supporting new material handling equipment (processes), and improving space utilization within unit load containers so that most of them can be efficiently utilized.

After the energy sector, transportation, and logistics is the most visible industry in the global effort to reduce carbon emissions. To reduce harmful environmental impact, there is a need to adopt more alternative energy sources and hybrid vehicles for over-the-road transportation and for material handling supply chain equipment inside facilities. By 2025, the transportation industry should have new vehicles category that are lighter, use less energy and run on alternative forms of energy than usually they are.

Buildings certified by the U.S. Green Building Council's Leadership in Energy and Environmental Design (LEED) program should be built and renovated for manufacturing and distribution. LEED certification should include scoring for equipment choices based on the types of services they provide (specific factors for distribution centers rather than, for example, hospitals). If the food logistics industry can do its role in achieving these goals, it will lead to the food industry's efforts to accomplish the same goals on a much larger scale.

Annie's Homegrown Inc. Embraces Sustainability on Multiple Fronts related to sustainability

Annie's Homegrown Inc., a natural food company, partners with more than 15 organizations dedicated to improving the lives of others and the earth's natural resources including non-depletion of ozone layer. The company's sustainability initiatives include:

Advocating for more sustainable food policies

Annie's belongs to the Organic Trade Association (OTA), an organization which is committed to promoting and protecting the growth of organic to benefit farmers, the environment, the public and the economy.

Expanding organic grains production

Through the Organic Grain Collaboration, the company partners with similar interest companies like Organic Valley, Stonyfield, Clif Bar, and others to address supply chain challenges facing the organic grain industry.

Actively engaging in industry groups

Bob Kaake, Annie's executive vice president and chief innovation and quality officer, serves on the board of The Organic Center, while Shauna Sadowski, the vice president of sustainability and industry relations, is on the board of the Sustainable Food Trade Association (SFTA).

Foundation Announces $90 Million in Grants for Sustainable Supply Chains in the Industry

The Gordon and Betty Moore Foundation' announced more than $90 million in grants designed to advance conservation by decoupling food production from deleterious environmental impacts.

"The anticipated global population growth and related demand for richer diets, including more protein, sugars, and oils, will place unprecedented pressures on our natural resources," says Aileen Lee, program director for the conservation and markets strategies and incoming chief program officer for environmental conservation at the Foundation. "Now more than ever, the need to produce food without causing severe environmental degradation is essential to maintaining the productive capacity and integrity of the earth's ecosystems."

The Foundation cited the following concerns in the food industry:

- Agriculture now covers nearly 40% of the world's ice-free land which is useful for farming, uses 70 to 90% of all extracted freshwater and comprises nearly 30% of global greenhouse gas emissions.
- Commercial agriculture is responsible for half of all global forest degradation.

- Despite concerted efforts to allow target fish stocks to rebound, overexploitation remains usual in global fisheries.
- Global aquaculture has more than doubled since the mid-1990s, bringing ecological consequences including the physical conversion of coastal ecosystems (e.g., salt marshes and mangroves), pollution of coastal waters with essential nutrients and antibiotics, and reliance on wild capture fish for feeds.

4.8. KEY SUSTAINABILITY CHALLENGES FOR THE FOOD INDUSTRY

Visibility

A key sustainability challenge is improving visibility and trackability throughout the supply chain process. The Sustainability Consortium's 2016 Impact Report found that the majority of consumer goods manufacturers lack visibility into the sustainability performance of their supply chains. This lack of visibility is a key barrier to improving sustainability for the industry. The consortium is administered by *Arizona State University* and the *University of Arkansas*, with support from *Wageningen University* in the Netherlands. Food and agriculture products in the industry have the largest share of impacts upstream (typically on-farm) where visibility is extremely low, the consortium noted. Only 20% of survey respondents collected data on fertilizer usage, greenhouse gas emissions and soil erosion. This research represents a significant potential risk to environment, but also an opportunity to improve both cost and sustainability for the supply chain process. Consortium data also suggested that in the berry, apple, and grape categories, nearly two-thirds of survey respondents could determine whether their suppliers had sourced crops from farms with zero conversion of high conservation value or high carbon stock lands in the past 5 years. Conversely, less than one quarter of pork, chicken, beef, dairy, and egg suppliers could make the same determination for their feed supply. Less than half of the companies purchasing palm oil could report what percentage of their supply was certified by the Malaysia-based roundtable on *Sustainable Palm Oil (RSPO)* or an equivalent organization; and even among those that could, less than half purchase 100% certified palm oil (Figure 4.7).

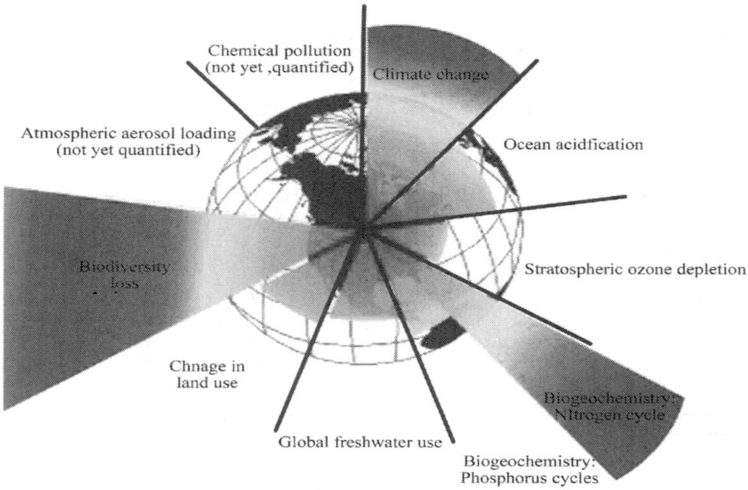

Figure 4.7: Climate challenges.

Source: https://www.thesolutionsjournal.com/wp-content

The need to improve supply chain visibility has resulted in a variety of initiatives for the food industry. Growers, processors, and packagers have adopted various metrics for improving sustainability. The regulatory community, responding to consumer concerns about the environment, nutrition, safety, and social justice, has created even more challenges for achieving sustainability for the food industry.

New Challenge: Genetically Modified Organisms

Last year, genetically modified organisms (GMOs) emerged as one of the most contentious food industry issues as a slew of retailers and manufacturers decided to label products as non-GMO in response to consumer concerns about the safety of GMOs. The voluntary labeling in itself created controversy within the food industry as some companies found this as an appeasement to bad science, while others saw it as addressing consumer concerns. Still others viewed the actions as responding to both existing and pending regulations. Pamela Bailey, president, and CEO of the Washington, D.C.-based *Grocery Manufacturers Association (GMA)*, told the GMA Science Forum in April that the GMO label rule in Vermont is the impact of non-science-based policies adopted. Bailey has also told the forum that studies show genetically engineered products are safe, and consumers have more confidence in the science of food safety than in the science of food nutrition. In response to these and other concerns, the GMA's *"Smart Label"* enables consumers to get information about products by scanning a barcode or doing an online search to get information on ingredients. The Smart Label grew out of recognition that the product information today's consumers are looking for exceeds what can fit on a package label and that leads to increment in packaging (Figure 4.8).

Figure 4.8: Genetically modified organisms (GMOs).

Source: genetically modified organisms (GMOs)

Expanding Population Challenges Sustainability Efforts for the Food Industry

The challenge facing the food industry increases every year due to expanding global trade, increasing limitations on agriculture, climate change and a growing population. The World Resources Institute refers to the global challenge of creating a sustainable future as *"The Great Balancing Act."* It's the 2013 World Resources Report identified three environmental impacts: ecosystems, climate, and most important water. For the world's agricultural system to achieve the *"great balancing act,"* the next four decades must match previous achievements in food production growth without converting ecosystems and increasing water usage. Few other factors, from climate change to urbanization to a lack of investment, will also make it hard to produce enough food as per the population growth. There is strong academic consensus that climate change–driven water scarcity, rising global temperatures, and extreme weather changes will have severe long-term effects on crop yields that will affect sustainability for the food industry.

Other Sustainability Issue

Other sustainability issues for the industry are making headlines, such as slavery in the supply chain and the use of antibiotics. Regulations and/or lawsuits could motivate or require food packagers to include labels that address these issues. In March, this year a judge found no support for a claim that *The Hershey Company and Nestlé USA* made actionable misrepresentations about supply chain slavery. Companies in the *United Kingdom*, however, are required to publish a slavery and human trafficking statement on their websites. As for antibiotics, the *American Association of Avian Pathologists* published a statement saying antibiotics should remain *"a viable option when appropriate and necessary for the health and well-being of the animal, even when marketing and consumer preference dictate otherwise."* The association claims that by removing antibiotics could result in situations where farmers are reluctant to allow treatment of flocks in order to maintain their antibiotic-free status. On the path to progress to sustainability, controversy should be expected and not something that necessarily undermines progress for the food industry.

4.9. TACKLING THE CHALLENGES

Looking at the growing body of evidence suggests solving food waste issue at first could help resolve many of the sustainability challenges faced by the food industry. Different studies in the world are linking food losses and waste to food inflation, food security, resource inputs and climate change. The global food industry and supply chain is currently experiencing its third bout of food inflation in 5 years because of poor agricultural harvests in the US, Russia, and South America. Analysts in the countries predict the average basket of food prices will rise by 15% in the years on years. The rapid hike in food prices is raising stupendous concerns about food security and its political and social repercussions. Food shortages in terms of production and price hikes have previously been responsible for riots in developing countries. With the projection of global population to rise to 10 billion in 2050 and resources becoming increasingly strained, there is a growing realization that raising production levels alone will not solve the problems facing the food industry. As it will be shown at the Sustainable Foods Summits in future, greater efficiency in supply chains can raise food output and help reduce consumer prices.

In a paper at the sustainability summit, the Food & Agriculture Organization of the United Nations (FAO) will show how a third of the food produced for human consumption is lost or wasted in the supply chain. Inadequate food storage and distribution the supply chain is the major cause of losses in developing countries, whilst waste at retail and consumer level is most responsible in affluent countries like US, UK, and Canada. A latest report by the Natural Resources Defense Council estimates that food wastage in the US amounts to $165 billion per year. The cost to the average American family is $2,275 per year. At the same time, over 15% of the population struggles to find enough food.

Keeping the social and economic implications of food waste apart, there are environmental impacts too. About 121 billion pounds (54.9 million metric tons) of food is wasted in the US, with two-thirds going to landfill. Food wastage in landfill is a major contributor to global warming, creating almost a quarter of methane gases. Agriculture also requires resources that are becoming increasingly scarce; uneaten food takes up 25% of the freshwater and 4% of the oil used in the US.

Today, by going through the importance of tackling food wastage, US food manufacturers, retailers, and restaurant operators formed The Food Waste Reduction Alliance (FWRA) last year. This alliance will look into reducing the amount of food waste going to landfill and increase food donations to hungry people. At the Sustainable Foods Summit, FWRA will state the growing importance of a collaborative approach to reduce food waste: how the private sector needs to work with the government and NGOs. Landfill diversion the major use of unused food is the focus of most retailers and foodservice operators. A growing number of such enterprises in the industry are setting up food recovery programs whereby food is diverted from landfill to hunger relief agencies, and/or for use in animal feed, composting or industrial applications. Feeding America through this method operates the largest food recovery program, operating over 200 food banks that serve almost 40 million Americans a year.

Today, Retailers are also becoming more aware of the economic benefits; reducing waste means lower hauling costs to landfill. Wholesale food market the largest chain of natural food shops in North America, is successful with composting food waste; over 75% of its stores have set up composting programs. Some retailers in the market have gone further by adopting a zero-waste policy. Four Albertson stores in California, US have achieved zero waste classification whereby 95% of their waste is diverted from landfill. Foodservice operators in the country like the *Bon Appétit Management Company* and *Sodexo* have also introduced programs to reduce waste. A major challenge in this, however, is changing consumer behavior, especially since about 44% of food waste comes from households. The role of food companies in the industry and retailers to encourage responsible consumption will be discussed at the *Sustainable Foods Summit.* Such methods in the process include educating consumers on how to read food product labels, shop wisely, and use food more efficiently (Figure 4.9).

Figure 4.9: Food wastage challenge.

Source: https://www.usda.gov/sites/default/files/styles/featured_image/public/usda blog

Tackling Food Waste is a focal theme of the 3rd North American edition of the Sustainable Foods Summit, the summit will bring together leading organizations involved in tackling food losses and waste; papers will be given by *FAO United Nations, Natural Resources Defense Council, Food Waste Reduction Alliance, Whole Foods Market, Bon Appétit Management Company, Feeding America*, among others. The biggest assets for companies looking to reduce supply chain complexity and food safety issues hampering the sustainability in the world, and become more sustainable and transparent in the process, are visibility and collaboration. Food companies worldwide, whether directly or indirectly, work with many levels of the value chain and as such have a voice in every step of production. Teams can't improve processes, protect against new risks and ensure they're providing quality, safe, and sustainable products if they don't have strong oversight and visibility into the performance of tier 1, 2 and 3 suppliers in countries. Issues inevitably arise when there's no insight into what's going on across the chain. Building strong supplier relationships in the supply chain and incentivizing the supply base to adopt sustainable practices is key for delivering safe, healthy sources of food and avoiding major outbreaks.

By treating all the suppliers as an extension of the business and raising their profiles as collaborators and partners, rather than vendors, food manufacturers are better positioned to find and fix weak links in the supply chain. Working together in the supply chain more frequently on process improvement, proactively sharing sustainability best practices, and building joint KPIs into contracts means there are stronger and more frequent lines of defense against safety vulnerabilities. These practices in the industry should be implemented on a global scale, not just on a regional or local level, and across every aspect of the food supply chain to give companies a fighting chance of eradicating safety issues before they occur. When it comes to the supply chain in the food industry, ignorance is not bliss. It's undoubtedly important to proactively monitor and take measures to reduce risk and protect the global food supply chain, because the end-product is not just something consumers buy and use, but something that impacts health. Consumers have a right to know the ingredients, what's in their food, and it is procurement's responsibility to make sure everyone—suppliers, manufacturers, distributors, and consumers—is set up for success by choosing to work with qualified, sustainable suppliers willing to proactively implement ethical, good sourcing practices into their own frameworks.

4.10. CONCLUSION

As per the discussion earlier, many of the options for food businesses and potential routes to more sustainable food supply chains discussed can create benefits both for the businesses involved and for others in the supply chain—from farmers to consumers the end user. Although creating niche markets for more sustainable locally produced products, conserved foods, and some manufactured foods can result in significant social, environmental, and economic benefits for those involved, greater overall gains could undoubtedly be made by improving the sustainability of mainstream agriculture and international supply chains. Various stakeholder initiatives, where food businesses work together with farmers, academics, innovators, governments, private firms, and NGOs, are important for raising the baseline for near-commodity and commodity food supply chains, including those important for manufactured foods. Definitely, by understanding the need for change within the food system, and the principles that would make a sustainable food system, is only the beginning of the journey. Innovative action to attain the sustainability is now needed by all parts of the sector's value network in all parts of the world – not just in Europe, but also in China, India, Africa, Latin America and South East Asia – to act on these opportunity areas for innovation and start to create a global sustainable food system. In order to reach towards the sustainability for the food industry all players are required to collaborate to curb on the hinders affecting the roadmap to goal.

REFERENCES

1. Anon, (2018). [online] Available at: https://www.ncbi.nlm.nih.gov/pmc/articles/PMC2610113/ [Accessed 12 Mar. 2018].
2. Anon, (2018). Available at: https://www.researchgate.net/publication/235965581 Sustainable benchmarking of supply chains the case of the food industry [Accessed 09 Mar. 2018].
3. Chain point, (2018). *Sustainable Food and Beverage Supply Chains, Traceability, and Visibility*. [online] Available at: https://www.chainpoint.com/industries/sustainable-food-beverage/ [Accessed 12 Mar. 2018].
4. Food logistics, (2018). *Sustainable Food: The Supply Chain Holds the Key to Success*. [online] Available at: https://www.foodlogistics.com/sustainability/article/12215449/sustainable-food-the-supply-chain-holds-the-key-to-success [Accessed 10 Mar. 2018].
5. Food Processing and Quality Control, (2007). [eBook] Available at: http://documents.wfp.org/stellent/groups/procumeeting_content/documents/events/wfp126925.pdf [Accessed 17 Mar. 2018].
6. Ifc.org, (2018). [online] Available at: https://www.ifc.org/wps/wcm/connect/30e5f0804b65403599 67dd08bc54e20b/The+India+Food+Report+2016_IFC+Chapter_web_.pdf?MOD=AJPERES [Accessed 12 Mar. 2018].
7. Keeping Food Safe, Sustainable, and Socially Responsible, (2013). [eBook] Available at: https://www.bsigroup.com/LocalFiles/EN-AU/Food%20Documents/BSI%20ANZ%20Food%20Portfolio%20Brochure_online.pdf [Accessed 17 Mar. 2018].
8. Rajendran, M. (2011) *Concepts of Food Safety and Quality Management Systems*. [eBook] Available at: http://old.fssai.gov.in/Portals/0/Training_Manual/Presentation%20on%20concepts%20of%20Food%20Safety%20and%20Quality%20Management%20Systems%20by%20Mrs.%20Malini%20Rajendran.pdf [Accessed 17 Mar. 2018].
9. Yamina, F., Mebarek, D., & Noureddine, B., (2014). *Food Risk Management and Sustainable Development*. [eBook] Available at: https://file.scirp.org/pdf/JSSM_2014043010065191.pdf [Accessed 17 Mar. 2018].

CHAPTER 5

FOOD INDUSTRY SUSTAINABILITY STRATEGIES

"To reverse the effects of civilization would destroy the dreams of a lot of people. There's no way around it. We can talk all we want about sustainability, but there's a sense in which it doesn't matter that these people's dreams are based on, embedded in, intertwined with, and formed by an inherently destructive economic and social system. Their dreams are still their dreams. What right do I -- or does anyone else -- have to destroy them. At the same time, what right do they have to destroy the world?"

—Derrick Jensen, Endgame, Vol. 1: The Problem of Civilization

CONTENTS

Sustainability is known to be a growing trend, which is extremely important for the organizations to make them aware about the concept and apply it in their daily activities. There is a triple bottom line of sustainability which includes social, environmental, and economic impacts. The competent multinational manufacturers of food product and food ingredients have the procedures for operations and strategic objectives and goals in place for working for sustainable production of food. However, apart from these, there are many challenges which make the manufacturing unit of specialty food ingredients to have solutions and equipments for improving the efficiency of resources all through the food supply chain.

5.1. INTRODUCTION

Sustainable Development is a policy endorsed by the government which embraces all the sectors of United Kingdom industry that ranges between heavy industry and food trade and manufacturing along with the civil society in beginning at entry level. The organizations which build and develop the actions of business with sustainability as their core ingredient may have a high probability to achieve success and excellence than the organizations which may need to change their actions. Consuming sustainably has now become a primary goal and most significant element of change and transformation for the consumers who are willing to contribute and become a participant in saving their land and planet. In response to this, the companies have designed their introduction of product and brand corresponding to the sustainable wellbeing and interests of the consumers. The main goal or framework purpose of sustainable development involves making all the individuals in the world satisfied with their fundamental and primary needs as well as enjoy and lad a quality life which excludes compromising of future generation's quality of life. All these principles help in forming and developing a base for the governmental policies all across the globe. In order to sustain a policy, it should take care of all the principles and apply it to food industries for achieving the sustainability.

5.2. THE ISSUE OF SUSTAINABILITY

The issue of sustainability and protecting the food quantity and quality is the primary strategic question for the stakeholders who a part of food supply chain like distribution are, packing, manufacturing of product, manufacturing of ingredients and agriculture, to the end users (customers).

There are many environmental responsibilities and obligations in food industry for preserving biodiversity, lower the waste of water and emission gases and taking care of livelihood with respect to wealth and health of local societies. Additionally, the food industry all over the globe experiences a worldwide increase in pressure related to food production, ingredient sourcing and raw materials for feeding the increasing population and that too in the environment which is consistently optimizing and supply chain control. In order to feed the present daily needs of the global population, we require and need yearly production of nutrients for delivering approx. 360 million tons of carbohydrates, 195 million tons of fat, 364 million tons of protein minimum. The fishery and agricultural industries are expanding and the projected rise in population may also expect clever and smart strategies (Figure 5.1).

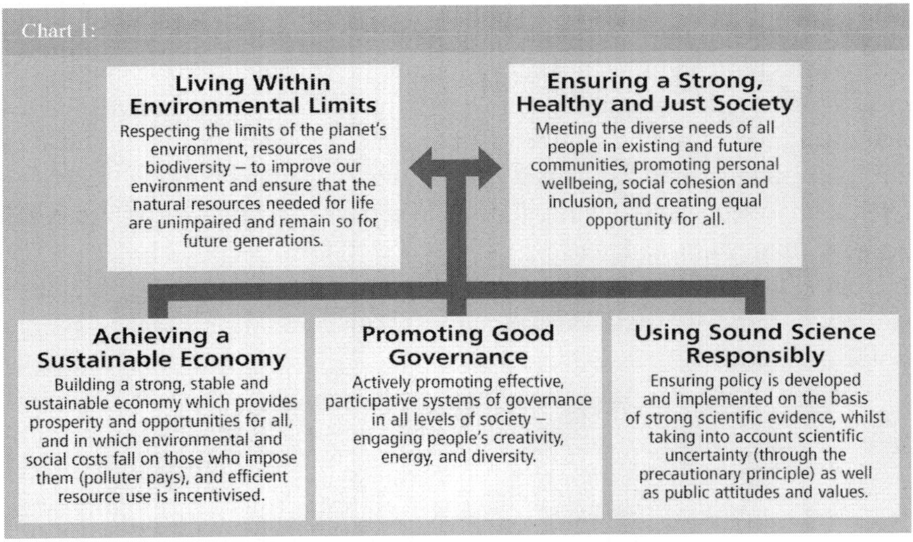

Figure 5.1: Shared UK principles for sustainable development purpose.

Source: https://www.gov.uk/government/uploads/system/uploads/attachment_data/file/69283/pb11649-fiss2006–060411.pdf

Dynamic and changing capabilities and sustainable management of supply chain are comparatively young fields to research which evaluates fast changing environments of industries and corporates. The most common example of fast-changing environment is the food industry. The customers keep with them high hopes for the safety of food as well as an increasing demand and need for the food which is sustainably produced and manufactured. The organizations who are completing these demands choose a segment of customer having high awareness and knowledge of all dimensions and scopes of sustainability: social, ecological, and economical situations where the food is manufactured and provided.

5.3. TOP TRENDS AND SUSTAINABLE PRACTICES IN FOOD INDUSTRIES

The objectives in sustainable development regarding food industry involve four primary objectives: promotion of sustainable production, improving nutrition of food, achieving the security in food, and ending the hunger.

5.3.1. Traceability

Till now, the product labels were providing information regarding the sourced ingredients which simply ignored the outcome of process of cooking. Huge privately and publicly structured data infrastructure are built near medicinal value and number of drugs and the function of lifestyle, diet, and food are hugely undervalued. The ultimate objective is to get the largest level of knowledge and security in food, which are driving over applications, analysis framework and sensors like Scio and Tell-spec which detects and identifies the existence of particular nutrients like gluten in Nima Sensor as well as build the data infrastructure for developing a global common footprint (Figure 5.2).

Figure 5.2: Sustainability in food supply chain.

Source: https://www.rit.edu/affiliate/nysp2i/infohub/supply-chain-sustainability/business-benefits

5.3.2. Maximization of Impact of Food on Health

Food is highly linked with the environment of health. The probiotic drinks and functional foods are being cured with the help of edible solutions like different pathogens which are emerging at a high pace. It is also known as trend of bio-informatics as well as the objective to solve food sources which provide scientifically, sustainable, and naturally proven solutions to health. Designing a new framework for food plays a significant role to minimize the consumption of food by simultaneously increasing the nutrient injection.

5.3.3. Reusing Bi-product and Waste

Waste is considered as a top problem while talking or discussing about sustainability of food. Exclusively in the European Union, approx. 90 million tons of food in a year are wasted and spoiled. There are huge opportunities for business in this respective area and many organizations are collectively gathering for making their individual contribution in order to solve this issue (Figure 5.3).

- **Food Cowboy:** Connecting and linking transporters and American Producers of Food with charity for lowering waste.
- **Food-Cloud:** Similar to Food Cowboy, but operating in Dublin have saved 1200 tons of food at present.
- **Last Minute Sotto Casa:** It is giving retailers a chance in Italy to give the products which are expiring to the customers who may then save their expenses.
- **Bio Bean:** Bio Bean is making the waste into a useful material by recycling it from coffee grounds to an advanced and contemporary biochemical and biofuels.

Figure 5.3: Sustainability.

Source: https://www.rabobank.com/en/about-rabobank/in-society/sustainability/food-supply/index.html

- **Food Fabrication:** 3D food printing varies from the technologies used in food manufacturing and is robotics based that are designed for automating the manual processes for huge production. It integrates the techniques of digital gastronomy and technology for manufacturing totally customized products of food. It introduces numerous opportunities like streamlining the cooking processes to production of new shapes which is not possible without the help of these techniques to accelerating the precision opportunities for nutrition.

5.3.4. Nerd Farming

Technology has essential power for enabling new opportunities in farming in interior warehouse settings. The production of food is retooled for accommodating and supporting urban living of high density and also maintains and sustains security of food in spite of future of increasing vulnerabilities instability of climate. Companies like Open Agriculture Initiative play around this region. The mission involves creating and developing more farmers for food production by generating software and hardware source stages for aero-phonic agricultural framework and sensor controlled hydroponic agricultural framework.

5.4. FOODS WITH LOW IMPACT ON ENVIRONMENT

As the population is increasing and has now reached 9 billion people, the industry of food has the pressure of finding new and eco-friendly solutions to food like edible insects which consist of high quality of amino acids, vitamins, and proteins for humans with lesser impact over the environment. Also, there are various companies working in this platform (Figure 5.4).

- **Entonote:** It is spreading the awareness on this topic of impact of food on environment.
- **EntoCube:** It is offering technology to insect protein ingredients and farm insects for food industry.
- **Bugsolutely:** It produces pasta which is cricket based.
- **The Algae Factory:** it is turning or transforming microalgae into sustainable and appealing food items like breakfast flakes and chocolate bars.

Figure 5.4: Logos for sustainability.

Source: http://www.ravenshoepackaging.com/blog/sustainability-in-packaging-is-this-just-politics/

In the year 2014, it was identified by an expert panel research of food industry that almost 100 industry executives gave ratings for general metrics for performance to "grow profit levels," "improve satisfaction of consumer" and "raise share in market" as their primarily important goals to be achieved in subsequent three years. It is clear that 70% of the executives oversee the long term of organization's survival and they also understand that there is a responsibility towards them and their company to protect and care for the environment where they execute their responsibilities.

5.5. LOYALTY IN FOOD SUPPLY CHAIN

In all the European customers, many regional differences are there with respect to loyalty in supply chain. The aggregate outcome from United Kingdom customers observed a stagnant opinions and suggestions between 2013 and 2014 as 64% of all the defendants called themselves as customers who are sustainability conscious, while 36% of them did not even care about it. Out of all the players from global industry, there are few significant issues and challenges that are upraised by the ingredient manufacturers which is associated to biomass management, cleaning, recycling, expenditure, water, and processing either through biofuel options or discovering the opportunities of new by-products.

The lifecycle assessment programs designed for ingredients and top agricultural raw materials through volume are in consideration for primary suppliers for ingredients and R&D investments to give more solutions for sustaining food for future generation. For many years, the largest beverage and food producers in the world have mainly focused over the supply side of raw materials in their sustainability events and programs for reaching newer goals. It can be a value addition to the present typical objectives with respect to reduction in packaging and greenhouse gas. Most of the multinational companies are in the support and helping programs of local communities of farming with respect to supply of goods, making consultations over issues like cultivation, fertilization, crop rotations and seeds selections, the handling and collection of crops as well as opportunities of microloans. The communities of farming mainly supported grow commodity crops like vegetables, foods, oil palms, cocoa, and coffee. All these programs and communities are collaborating with manufacturers and other interest corporations.

5.6. SUSTAINABILITY AS INNOVATION

The conception of sustainability has grown and developed across three-time periods. Sustainability was initially observed as an operational issue which consisted mainly of protective labors for lowering environmental footprints of the company. It moved from reduction of cost to uniqueness, also known as Sustainability. At present, the world is said to be in the middle of third decade where the concept of how a business must be done is of utmost importance (Chouinard, Ellison, and Ridgeway, 2011) (Figure 5.5).

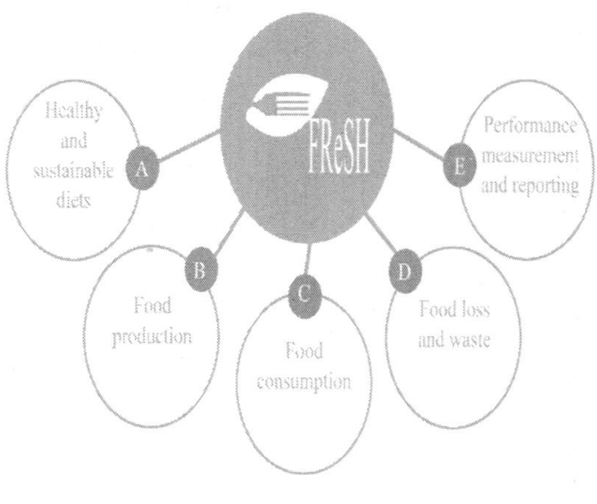

Figure 5.5: Food sustainability solutions.

Source: http://www.ravenshoepackaging.com/blog/sustainability-in-packaging-is-this-just-politics/

5.6.1. Phases in Innovation of Sustainability

The initial phase in the process of innovation involves viewing compliance as a chance and opportunity that includes working with different organizations for implementation of unique and creative explanations as well as own the ability for shaping the regulations and roles. It allows the organizations to try testing with processes, materials, and sustainable technologies.

The second phase involves bringing sustainability in the value chains. The main abilities required in this involve know-how and skills in the techniques like managing the carbon and assessing life-cycle, the capability for redesigning operations for using lesser amount of water and energy, generate lesser emission and waste along with the capacity for making sure that traders and suppliers are able to perform eco-friendly operations.

The third stage involves designing and framing sustainable services and products for becoming environment-friendly. The capabilities required to complete this phase include the skills and capabilities to identify the goods and services that are not environment-friendly, the ability for developing original support for public for sustainable contributions as well as managing the expertise to increase supplies and manufacturing of green materials and products.

The fourth phase in sustainable development involves developing new models for business which may capture and provide the values that alter the fundamentals of competition. It can be attained by understanding the demands and wants of customers as well as assess various ways for achieving these needs.

The fifth and the last phase involves developing next practice platforms. The skills needed for effectively completing this stage is know-how and expertise of the way in which non-renewable and renewable resources can impact the industries and ecosystems of the business as well as expertise for creating and developing regulations, technologies, and business models in various industries (Nidumolu, Prahalad, and Rangaswami, 2009).

5.7. TOP SUSTAINABILITY ISSUES IN FOOD INDUSTRY

- Responsible Management of Supply Chain;
- Agricultural and Industrial consumption of water and prevention of pollution;
- Immigrant/seasonal/local labor rights;
- Sustainability reporting;
- Sustainable plans for food and farm.

Consumers are highly interested to learn about a broad category of sustainability strategies and actions that are taken by the companies. However, most of the companies which are patronized by the customers depict a high level and intensity of commitment and responsibility (Figure 5.6).

Figure 5.6: Sustainability and its significance.

Source: https://www.rit.edu/affiliate/nysp2i/infohub/supply-chain-sustainability/business-benefits

Unfortunately, there is a small percentage and proportion of consumers who presently believe and have faith that the corporations perform in a way which helps in protection of environment. However, the consumers also believe that volunteer groups (41%), Non-profit organizations (42%) and individuals (44%) are achieving excellence in this domain. More than half of the customers have the opinion that companies must be doing a lot more for protection of environment, which clearly indicates that there is a space for the organizations to increase the efforts.

It is authoritative that organizations be transparent, straightforward, and clear as they move ahead. Only some of the consumers trust the companies of food and beverages for following the actions which promote and protect the sustainability in environment. Engagement of consumers in sustainability increase consistently. For this, most of the consumers have a vision in organizations for good leadership, but they are also attached with these companies due to the sustainability efforts. In current years, the companies in food and beverage have significantly invested in communications and initiatives of corporate social responsibility. Many consumers have this notion with them that companies' actions and decisions are equally important as the products which they are selling. Thus, the efforts of sustainability are a significant part of Corporate Social Responsibility with societal and social initiatives (Table 5.1).

Table 5.1: Social, Environmental or Ethical Issues

Percentage of U.S. consumers interested in learning more about what companies are doing to address this issue	
Treat animals humanely	64%
Reduce amount of waste produced	61%
Use more recycled content in products/packaging	46%
Reduce food waste	43%
Use less packaging	43%
Increase use of renewable energy	43%
Reduce energy consumption	41%
Prevent global warming/reduce greenhouse gases	40%

Source: NMI's Sustainability Consumer Trends Database

There are various chances for increasing business of food and beverages, however maintaining the position along with the dynamic needs of consumers, employing strategies to acquire and key technologies, leveraging new packaging and understanding idealistic preferences are all necessary in the competitive marketplace.

5.7.1. Changing Preferences of Customers

Exotic and bold favors, wellness, health, and snacking are known to vanguard the preferences of consumer food, while the idealistic generation is driving the choice choices with adventurous tastes. Subscription home delivery of food will also evolve, by thinking of wishes of consumers to be a share of making process of meal with easy going ingredients that are fresh and healthy and recipes. Similarly, convenient, and nice packaging such as choices that are bowl based that features quinoa, kale, and many more are in demand because keen consumers observe for speedy and healthy eating choice. Being on top of moving preferences and lifestyles will be necessary for profitable and relevant companies of food and beverages. As it has been witnessed, big organizations in food packaging which are not in trend and line with strong snacking has caused lower sales.

5.7.2. Mergers and Acquisitions

Many of new lines of product and healthier ingredients and products will be added by most of the organizations in 2018 to the current food and drinks. However, instead of developing the new products in organization's current functions, a lot of organizations acquire the businesses which are developing type of ingredients and foods which they need. The acquisitions and mergers are done at lesser cost instead of spending and incurring cost and time in internal research and development for creating newer products. Acquisitions is considered as a new Research and development technique for most of the companies in food and beverages who want to increase the brand and offering of the product.

5.7.3. Sustainability and Safety of Food

In the current years, there is a lot of lag of consumer confidence for United States food supply safety. As per a current Food and Health Survey by International Council of Food Information, nearly 60% of people in America have confidence in the safety of supply of food which was 66% in the year 2016. Also, the manufacturers of food may mitigate the risk as well as remove the risk and also vanish the area. For most of the business, preventive efforts and food safety may consistently be on top of their mental state in the year 2018.

5.7.4. Cost Efficiency

Many consumers, especially idealistic desire to understand that the producers of food are charitable and green along with the increased products, control their water consumption, use alternative energies like solar power, and reduce wasteful packaging and more. Since these significant sustainable alternatives can be expensive for operations, the organizations should accelerate effective green framework and also be known to tax credits that save cost such as deduction without food deductions for helping in offsetting the significant efforts.

5.7.5. E-commerce

The Grocery store's shelf space has now become very competitive and there are various selling channels which are attractive and are taking a high on performance. At present, selling directly to the consumer sites as well as third-party platforms for e-commerce account for nearly 36% of the sale of food and beverage. With the acquiring of whole foods of Amazon, the demand and need for the delivery and subscribing of food services may consistently grow and increase in the year 2018, since the busy consumers prefer doorstep delivery of goods. The increasing competition for shelf space will lead to an increase in demand of home delivery. All the customers give orders for their products using their devices and the product is delivered to them in hours or is available to be picked up from any nearby store.

5.7.6. Technology

In the coming years, the customers are going to do a lot of research prior to buying a product and it includes beverages and food. By doing a speedy research for their smartphones, the consumers are able to make a quick and informed choice for their products which they intake. The information about the products which provide energy and protein boost required for workout, the drinks that promote and enhance a healthy glowing skin, the core history and values of the organization which is producing the product are all relevant for the customers. Successful business into food and beverages may use social and digital platform for connecting with the consumers by giving a robust avenue for product research, forums, stories, and tips and a lot more information to the customers who are hungry for the same. Harnessing the internet of things and big data are important parts of that overall strategy as well (Figure 5.7).

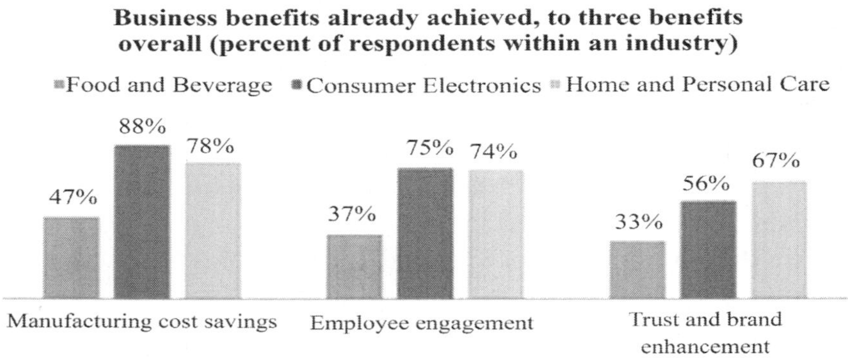

Figure 5.7: Business advantages already achieved (Percent of Respondents within an industry).

Source: http://www.sustainablebrands.com/news_and_views/products_design/cheryl_baldwin/food_industry_playing_catch_sustainability

Companies into food and beverages are initiating to understand the significance of sustainability with increasing companies who are working for advancing the social and environmental improvement of the product packaging, manufacturing, and ingredients all through the product lifecycle. Industrial segmentation of approx. 100 worldwide product organizations which included 34 beverage and food companies included in sustainability of product who were surveyed and questioned by Pure strategies disclosed that though food manufacturing sector is initiating to take the significant steps, a lot of strategic work is required for the industry for learning about the top performers of the market in all the industries.

5.8. PATHWAY TO PRODUCT SUSTAINABILITY

The value of business which was comprehended by Heinz reflects the high advantages as cited by organizations having active programs for product sustainability. As per the Pure Strategies' recent report, The Path to Product Sustainability, the list includes savings of manufacturing cost, followed by productivity and engagement of employee. Apart from the exceptions, the food industry follows other sector having only 47% of companies in food who were surveyed for achieving the savings in manufacturing expenses, in comparison to 86% of aggregate top performers of the survey. In all the companies who are into personal and home care and consumer electronics, almost 78% and 88% of the organizations have reported to secure and achieve positive results. The shortfall of food industry in this domain increased towards all the advantages like enhancement of brand and trust, engagement of employees and other probable gains (Figure 5.8).

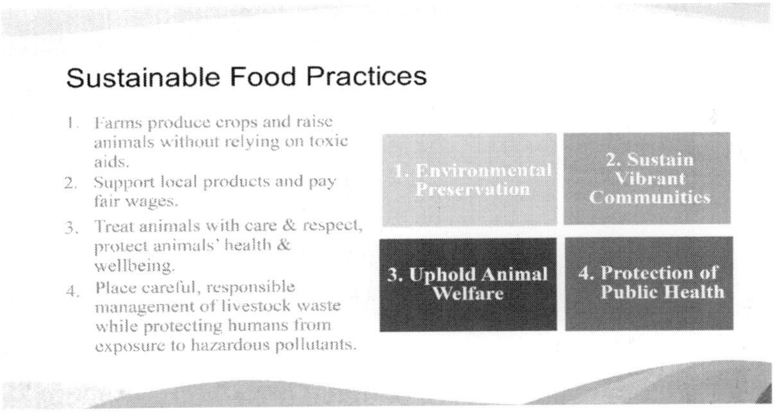

Figure 5.8: Sustainability food practices.

Source: https://poesiaemtranse.com/what-is-sustainable-food/

In the present climate and atmosphere, where desires and perception of consumers are getting thoughtful services to count with, the food and beverage corporate's sustainability is shifting from selection for processors for considering to be imperative. In a current survey that was related to Packaging Matters, the provider of packaging solutions West-rock identified 98% of consumers showed sustainable actions which are extremely significant for the view of CPG brand. Having more emphasis over the sustainability, the industry of food and beverage is replying with projects and programs who focus over prevention of pollution, conserving efforts, initiatives to recycle and more. In reality, food engineering is developing awards program for honoring plants having improved performance and activity in sustainability as well as individuals going beyond the compliance of environment. Some of the companies were winners of OpX Leadership Network's Sustainability Excellence in Manufacturing Awards, presented at Food Engineering's Food Automation & Manufacturing Conference last April (Figure 5.9).

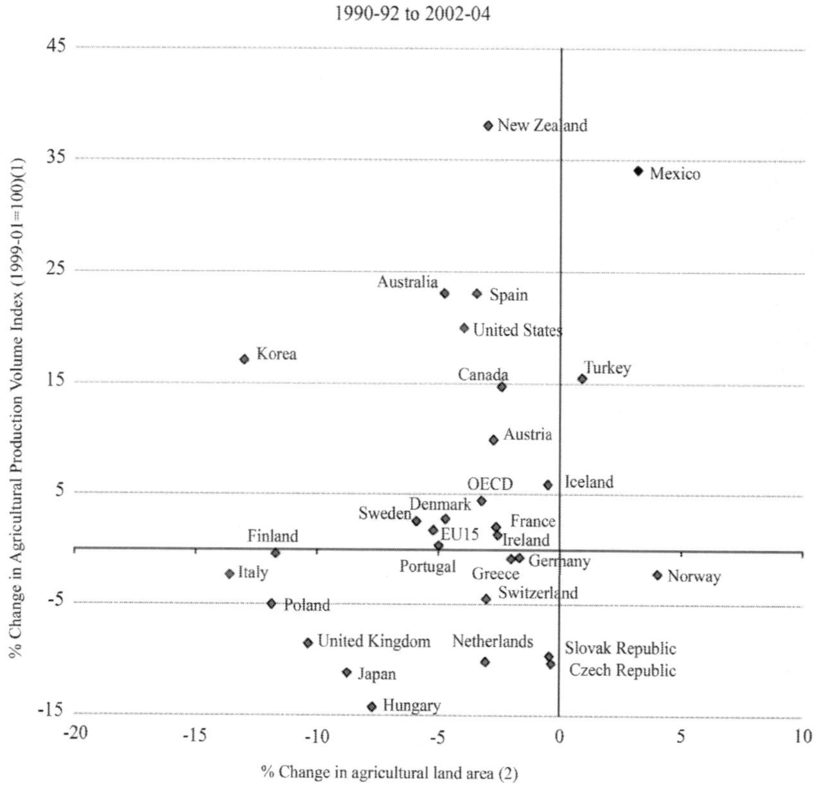

Figure 5.9: Volume index of production of agriculture and area of agricultural land.

Source: This figure is taken from OECD (2008), Environmental Performance of Agriculture in OECD countries since 1990, Figure 1.1.5, p. 46.

- Production index of agriculture is a volume index of aggregate livestock and crop production. The data in this picture are averages taken for the year 2002–2004, with 1999–2001 as its base year = 100.
- Slovak Republic and Czech Republic: Average 1990–1992 = Average 1993–1995.
- Luxembourg and Belgium are not included as the data available is from 2000 to 2004.
- Change in percentage in aggregate national land area for agriculture is depicted in 1000 hectares, 1990–92 to 2002–04.

5.9. INITIATIVES TAKEN BY COMPANIES: CASE OF WHITE-WAVE

Recently, White Wave Foods released the sustainability report, where it outlined the commitments and efforts for driving positive change as well as deliver over the "Changing the Way the World Eats for the Better" mission.

"Our industry faces challenges as the need for action on climate change, water scarcity and other sustainability issues becomes more urgent," the report states. "For us, addressing these challenges means having a clear sustainability strategy that guides our actions, drives solid goals, and motivates us to achieve them."

The strategy of the company has three main pillars. The first involves cultivating a receptive and responsive food framework that focuses over ways for improving efficiency, implantation of sustainability in products, form innovative designs as well as making an investment in sustainable agriculture along with best ingredients. The subsequent pillar involves illumination of better choices that demonstrated integrity and openness all through advancing the transparency along with focusing over healthy options for meeting the changing needs of the consumers. The third one involves empowering the employees of the organization by encouraging the engagement and meeting in sustainability as well as creating and developing positive change of the company, communities, and culture in which White wave works in.

According to Deanna Bratter, director of corporate sustainability for the company, "We have been measuring our environmental footprint since 2006. From 2006–2013, we reduced our GHG emissions 32%." "In just the past two years, White Wave has reduced waste to landfill by 52% by focusing on reducing waste and increasing reuse and recycling." (Table 5.2)

Table 5.2: Economic, Environmental, Social as Well as Science and Technology Tools and its Impacts

	Economic tools	**Environmental tools**	**Social tools**	**Science and technology tools**
Economic Impacts	Assessing the cost-benefit analysis	Regulations of environment for internalizing costs	Green practices to reduce poverty in agriculture	Alternative farm systems

Environmental Impacts	Environmental services payments	Mechanism of cross-compliance of environment	Provisioning social infrastructure and framework in the rural region	Training of skills in green practices of agriculture
Social Impacts	Supporting farm incomes	Producing goods and services related to the environment	Measures for structural adjustment	R&D of agricultural biotechnology
Technical Impacts	Tax credits of R&D for research related to agriculture	Trading systems and water charges	Green extension programs in rural regions	Public Research for the promotion of eco-efficient agriculture

5.10. GREEN GROWTH OF SUSTAINABLE FOOD INDUSTRY

Moving in the direction of green growth in the sector of food and agriculture involves a broad range and variety of policy events, few of them are specific to the sector while some of them are economy based. Actions that are taken for promoting green growth must complement and substitute the current priorities and consider dynamic priorities for being most economical in weakening environmental pressure and force along with providing and catering to the future needs for fiber and food as well as provisioning the ecosystem services. A strategy for green growth food agriculture and food needs a well-targeted, coherent, and coordinated answer to the social, environmental, and economic challenges and issues in the upcoming decades (OECD, 2007). This strategy is inclusive of five primary phases regarding policy actions (Figure 5.10).

- Outlining the business being common projections regarding social, environmental, and economic trends significant to the sector of food and agriculture.
- Developing and creating a long-term vision and idea of social, environmental, and economic objectives for food and agriculture (setting the targets that are easily measurable).
- Identification of priorities of policies and policy instruments which are cost-effective (for achieving the targets).
- Implementation of policies which includes involving the

stakeholders, private, and public partnerships and coordinating various ministries for ensuing cohesive implementation of policy.

- Controlling and monitoring the progress in the direction of green growth (Policy reviews and indicators).

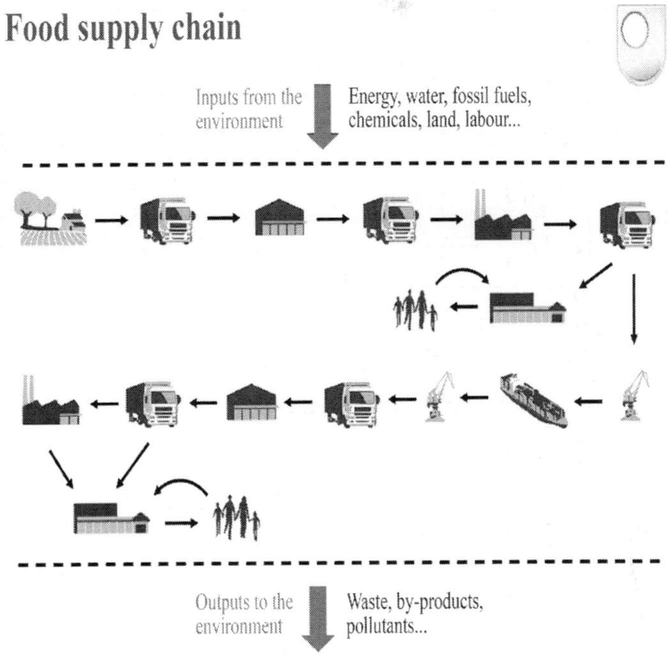

Figure 5.10: Food supply chain.

Source: http://eschooltoday.com/global-food-waste-and-food-loss/how-is-food-wasted-in-the-food-supply-chain.html

Therefore, the political economy needs to be right and correct along with the support from the aggregate policy environment, buy-in, ownership, and good governance from the general public, , and stakeholders. Putting and keeping the sector of food and agriculture over greener pathway of growth implies adjustment and change, that sequences the policies in long and short term, shouldering either the costs or reaping the advantages, not only with respect to flow of income but stocks of wealth that has distributional and social results (Figure 5.11).

Figure 5.11: The cycle of food supply chain.

Source: http://eschooltoday.com/global-food-waste-and-food-loss/how-is-food-wasted-in-the-food-supply-chain.html

5.10.1. Sustainable Programs

The sustainability Programs by Smithfield Foods is arranged and organized over five primary pillars: helping people and communities, food quality and safety, environment, animal care. The company currently released and unveiled its Report for Sustainability and Finance that highlighted few achievements like clarifying using antibiotics in farm animals that established an ingredient wordlist as well as making good progress in targets to lower natural resources. The report took into consideration that company was meeting its goal to reduce GHG emission and also met the target to reduce the use to energy four years after the deadline in 2020. "The sustainability program by Smithfield has triggered change all through the operations of the company in various methods" according to Stewart T. Leeth, vice president of regulatory affairs and chief sustainability officer for Smithfield Foods, Inc.

It also includes various systems for animal care management, procedures, and policies for ensuring proper and accurate treatment of dogs. As per Leeth, it can impact the relationship and reputation of Smithfield with consumers and also help the level of production of animals. "Healthy animals are more resistant to disease, gain weight faster, and healthy sows have larger and stronger litters," he says (Figure 5.12).

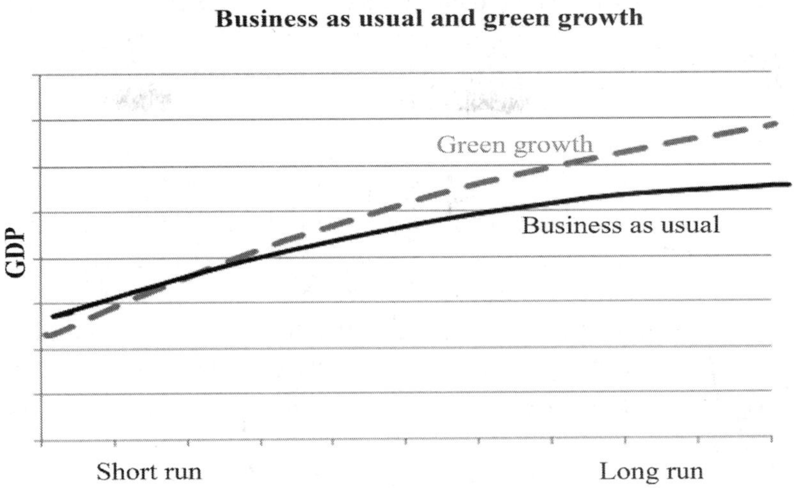

Figure 5.12: Business as usual and green growth.

Source: OECD Secretariat

In the year 2015, all the independent operating organizations were reorganized by Smithfield in a single corporate shed. Post this, the corporation optimized and adjusted its network of logistics for improving the efficiencies, lower complexities, reduce operating costs, reduce carbon footprint of company and also improve the services of the customers. Also, six Smithfield facilities have attained zero waste to landfill status and therefore no longer giving solid waste for incinerators or landfills. For increasing the employee engagement, White-Wave organized the annual Wave Programs for nominating the coworkers exemplifying value of companies.

In line with the strategies adopted by different companies, Big Heart Pet Foods has made efforts to motivate and encourage the workers for participating actively in the sustainability efforts. The corporation has a program and space where the employee in New York bring electronic waste in work for diverting dangerous materials and sources from the landfills. The company has also got third place for OpX 2015 Excellence in sustainability in manufacturing awards. The company is working to bring optimistic change in the surrounding community through donation of the recycling proceeds of the plant in various local organizations. The proceeds of scrap materials had gone to Gobike Buffalo that helps in developing biking infrastructure. The recycling proceeds of Wood Pallet were being sent to Re-Tree Western New York that helps in reforesting the local region. According to Michael Jackson, distribution supervisor for Big Heart Pet Foods, "The money allowed for additional tree plantings—turning scrap wood into saplings." Also, proceeds of paper recycling went to Erie and Buffalo County Libraries that helps turning paper scrap in new books. The proceeds of Electronic waste went to special times and good days of the camp, that benefits and are advantageous to children having cancer. The initiatives can be represented as reduction in benefits of plant waste not only the organization, but the community. The program got initiated to try to give tangible and direct advantages to the Buffalo community through the efforts of waste reduction. Jackson says, "to go from making an abstract impact on 'the environment' to making a direct and visible positive environmental impact on the city of Buffalo."

5.10.2. Potential Green Growth Opportunities

As per the United Nations, nearly one-fifth of the population of world dwell in areas of water scarce as well as another people face shortage of water. The significance of water becomes vibrant as a commodity.

"We have a strong water use reduction strategy at White Wave," says Bratter. "We work with World Wildlife Fund to assess water risk for our operations and key suppliers worldwide that allow us to better understand our impacts and opportunities." (Table 5.3)

Table 5.3: Potential Green Growth Policies

Environmental standards and regulations	Make improvisation in enforcement of certifications/ standards and regulations to a retail sector from farm gate. Strengthen the standards and rules for land management, soil quality and water. Enforce the monitoring over excessive usage of fertilizers and agrochemicals in production.
Supporting aids	Target public investment in green technologies. Target the outcomes of environment where possible, otherwise the practices of target production favorable for environment. Compensate provisioning of public goods of environment beyond the level of reference like drought/flood control, carbon sequestration and biodiversity).

Economic Instruments	Cater to constraints of policy like governance, conflict, etc. in the economies that are less developed. Facilitation of private investments in green management and technologies. Implementation of trading schemes for carbon emissions and water rights. Impose taxes and charges on using inputs that damage the environment. Price inputs for reflecting the shortage value of natural resources.
Trade Measures	Eliminate restrictions and export subsidies over agricultural products. Aid the well-functioning output and input markets. Non-tariff and lower tariff constraints over agriculture and food product having in mind a heavy impact over the concerns of environment like sustainable and biodiversity resource use.
R&D	Undertake private/public associations for the research of green agriculture. Promotion of private R&D in agriculture through tax and grants credits. Increase and enhance public research over the sustainable systems of food and agriculture.
Assistance in Development	Allocation of more funding and capital for aiding agriculture in trade projects. Increase the agricultural profile in strategies for reducing poverty. Allocation of more development help for sustainable initiatives for environment, in food and agriculture.
Advice, Training, Education, and Information	Incorporation of sustainable approach to train, educate, and advice the programmes all through the food chain.

White-Waves monitors, and measures use of water in all the company owned manufacturing plants. The present goal involves reducing the intensity of water and the gallons of water that is utilized per pound of good, approx. 20% till the end of 2015. The intensity of Water Company wide has also been lowered by 6% since the year 2013.

5.11. BEST PRACTICES TO INVEST IN

The food companies are making investments in sustainability of food, having no single organization in survey that expects a decrease and reduced budget and expense for the work and approx. 20% working with double-digit growth of budget in the area. It starts with charting a relevant and focused direction having meaningful objectives. The Pure strategies identified that food corporations have goals having 80% of goals surveyed and questioned noting objectives of product sustainability. For instance, the Coca Cola Company is working to achieve agriculture, climate protection, packaging targets over water which includes sustainably sourcing the significant key ingredients of agriculture by 2020.

There are many opportunities and chances for the food industry to make improvisation in utilizing the assessments as well as other tools for informing decisions of product as well as integrating the sustainability in packaging and product development process. Two primary strategies are currently adopted by many of the companies: Make investment in engaging suppliers as well as work for integration of sustainability in process of development.

5.11.1. Supplier Engagement

Engaging the supplier is considered to be most valuable and informational approach for sustainability-oriented innovation. The food industry, however, avoided to report to be as active and lively with engagement of the suppliers like other industries, which resulted in not getting much benefits. How can food companies ramp up activity and derive more value from supplier engagement? The way in which food companies are able to increase activities and get more value from engaging their suppliers involves initiating sourcing significant ingredients through sustenance by supporting, collaborating, and communicating the supply chain and suppliers.

After prioritizing the establishment of goals and objectives, the organizations are required to initiate sourcing significant ingredients through sustenance by supporting, collaborating, and communicating the supply chain and suppliers. Coca-Cola is closely working and associating with the suppliers for the advancement of the program as it understands and is well aware of the approach. The sustainable sourcing efforts of Coke involve sharing prospects for agriculture as well as a collection of information from the suppliers regarding actions that are taken for meeting the guidelines for identifying the best actions and motivating application all through the supply chain for helping in the achievement of goals of 2020 (Figure 5.13).

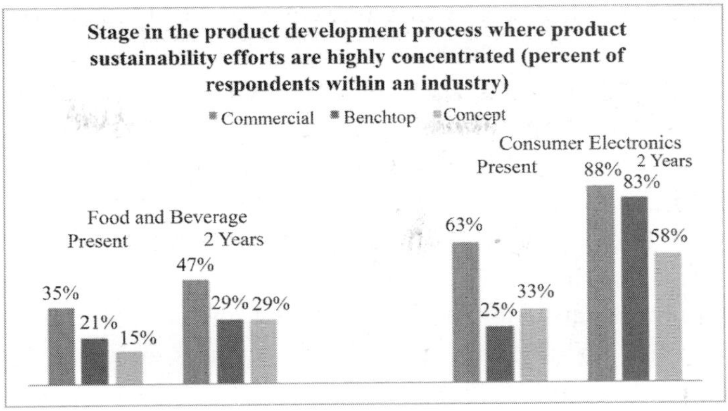

Figure 5.13: Stage in process of product development.

Source: http://www.sustainablebrands.com/news_and_views/products_design/cheryl_baldwin/food_industry_playing_catch_sustainability

5.11.2. Product Development

The most crucial area in which the food companies are required to comply is in the development process of packaging and product. Notably, the major organization in the aggregate market have started initiating to have plans for concentrating the efforts on product sustainability all through the process of development, particularly in the earlier phases of pilot development/ bench top. Approximately 29% of companies into consumer electronics concentrate over sustainability in phases of pilot phase/bench top, which may rise and increase to 83% in subsequent two years. Only 20% of companies in food industry do this presently and 30% have expectations in two years (Bradbear, Catie, and Friel, 2011). Coke involves sustainability in strategic platform of "Research and development," putting it in Center and front at the time of product development. It simply means that at every gate of stage, the managing the reviews of social attributes and environment products, completely integrating the advantages from concept of product development to commercialization.

5.11.3. Making Investment in Sustainable Food Industry

Since the food industry is going ahead, their investment is required to be strategic for yielding huge advantages. Getting appropriate actions of individuals who are liking and appreciating the best value of business may help the food corporations to comply (Polonsky, 2011). The combining of early and fast integration and supplier engagement, the strict objectives of every firm into food and beverage industry as well as the internal hold of sustainability of product in this sector can project the companies towards top position of the company. The companies will realize the advantages of the tangible business and will win more sustainable food that is present at local market area. There is an increase in consumers who will buy sustainable and ethical products such as animal-friendly, fair trade and organic products. For any other product which can be marketed, the consumer acceptance is critical for achieving success in sustainable products. A research conducted by Vermeir and Verbeke identified that all the young customers are a significant part in sustainable food consumption other than any sector. The consumers have more preference for the food products which are sustainable as they are better with respect to freshness, safety, quality, and taste. If the companies are able to make the products that are needed, the consumers have to purchase the products, irrespective of the fact that their prices are higher in comparison to the non-sustainable food product.

5.11.4. Transition and Transformation in Core Ideology

Food Organizations that are not developed with values of sustainability are required to make transition and transformation in core ideology as well as change the culture of organization with time and move towards sustainability, otherwise they may lose the chance of getting successful in the present world (Crittenden, Crittenden, Ferrell, Ferrell, and Pinney, 2011). Sustainable Food products show a small and minor fraction of share in the market and the industry into organic food is completely similar to this, where the organic food accounts to less than 3% of total sales of food in the year 2006 (Sheth, Sethia, and Srinivas, 2011) (Table 5.4).

As the environmental degradation is increasing it is of serious concern around the globe, while in future, the degradation of environment will consistently act as the biggest constraint for future production of food that may contribute to reduced affordability, quality, and quantity in most of the nations (Chabowski, Mena, and Gonzalez-Padron, 2011; Bradbear, Catieand Friel, 2011). The primary contributors for this degradation involve food waste, packaging, biodiversity, water waste and greenhouse gas emissions. The environmental efforts like energy, water, and land are utilized at every phase of food productions which includes consumption, retail, distribution, food packaging and processing and agricultural production.

Table 5.4: Agriculture and Green Growth: Illustrative Examples of Complementarities (+) and Differences (-)

	Economic contribution of agriculture to green growth	Environmental contribution of agriculture to green growth	Social contribution of agriculture to green growth
Economic contribution of green growth to agriculture	Agriculture as a driver of economic development while Green Growth can improve agricultural performance (+)	Green labels and payments for eco-services can contribute to economic returns in agriculture (+)	Higher skilled jobs and activities can diversify and contribute to rural development (+)
Environmental contribution of green growth to agriculture	Environmental measures may slow agricultural growth in the short-term and may result in … (-)	Green Growth will yield environmental co-benefits in agriculture through resource conservation and sustainable use (+)	Reform of support to relieve environmental stress and payments for ecosystem services can enhance farm incomes in rural areas (+)
Social contribution of green growth to agriculture	Green Growth may detract from efforts to improve food security in the short-term (-)	Green Growth will necessitate structural adjustment measures in transition periods (-)	Food security, poverty reduction, and rural development will be enhanced in the long-run through Green Growth (+)

5.12. CONCLUSION

Major companies in most of the sectors are accelerating the sustainability of product for delivering relevant social and environment change as well as for developing benefits for the business. In the year 2011, Heinz, a renowned brand had taken a strategic approach in manufacturing for achieving reduction in energy use and waste, absolute emissions with savings of money as well as strong engagement with the employees. The organization has achieved it using employee-based projects which identified and experienced efficiency solutions like modifications of processing lines which mixed cold and hot water for reducing the needs of reheating with development of cohesive renewable and conservation of energy efforts. It has been observed that supplies of fresh water are lowering because half of river of world are polluted and depleted seriously, damaging the production of food. The agricultural sectors which are known to be highly and extremely susceptible to the influence of change in climate are irrigated grain producers, beef, and sheep, which are followed by grain producers, beef, and sheep of dry land, and subsequently some fisheries near the world.

The extent to which the customers feel that the issues of the environment are difficult is linked to disagreement in timeframes, responses, and causes where the actions for remedy should be taken. In most of the cases, the customers do not bother to look for the solution to any problem as they have a notion that it is something which does not need to be care about, instead it can be taken care of in future. They have this notion because the issues of environment have no secured deadline that are set in place as well as no media promotion for making them inclined to perform the same.

REFERENCES

1. Anon, (2018). [online] Available at: https://www.ncbi.nlm.nih.gov/pmc/articles/PMC2610113/ [Accessed 12 Mar. 2018].
2. Favero, C., & Soriani, C., (2004). *Corporate Social Responsibility in the Food Industry*. Valuing the use of corporate social responsibility in the Australian food industry. [online] Dspace.unive.it. Available at: http://dspace.unive.it/bitstream/handle/10579/7442/840125–1173024.pdf?sequence=2 [Accessed 14 Mar. 2018].
3. Fritz, M., & Schiefer, G., (2008). *Sustainability in Food Networks*. [eBook] Available at: https://ageconsearch.umn.edu/bitstream/52648/2/fritz.pdf [Accessed 14 Mar. 2018].
4. Ifc.org, (2018). [online] Available at: https://www.ifc.org/wps/wcm/connect/30e5f0804b6540359967dd08bc54e20b/The+India+Food+Report+2016_IFC+Chapter_web_.pdf?MOD=AJPERES [Accessed 12 Mar. 2018].
5. Rajendran, M. (2011)*Concepts of Food Safety and Quality Management Systems*. [eBook] Available at: http://old.fssai.gov.in/Portals/0/Training_Manual/Presentation%20on%20concepts%20of%20Food%20Safety%20and%20Quality%20Management%20Systems%20by%20Mrs.%20Malini%20Rajendran.pdf [Accessed 17 Mar. 2018].
6. Yamina, F., Mebarek, D., & Noureddine, B., (2014). *Food Risk Management and Sustainable Development*. [eBook] Available at: https://file.scirp.org/pdf/JSSM_2014043010065191.pdf [Accessed 17 Mar. 2018].

CHAPTER 6

CONTROLLING THE SUSTAINABILITY OF FOOD SUPPLY CHAIN

"Since the explosion in ecommerce in recent years, the simple way to say it is that the Supply Chain and Logistics profession has become kind of cool"

—Scott MacKinnon, Plog

CONTENTS

As food industry is in growing phase, the unique challenges like issues of logistics, new supply routes, need dynamics and investment are also increasing and are redesigning the future of industry of food logistics. Therefore, it is a significant responsibility for the food industry with respect to managing the demand and sustaining the sources of food for the increasing population to be innovative. The sustainable supply chain in the food industry and logistics gives an essential and accessible guide for managing the food supply chain which considers food supply chain starting from the farm and ending to fork.

6.1. INTRODUCTION

Sustainability is discussed extensively all over the world. Its existence is there in all the stages and levels of business executions all across the supply chain. The fundamental base to design and run all the operations should originate from the need to achieve the criteria of sustainability. Caring for sustainability is not considered to be any operation conducted as voluntarily. All the companies must meet a lot of legal compliances for achieving the sustainability. In this chapter, the challenges prevailing in sustainability of food supply chain will be discussed. Concept of food Supply Chain involves a complicated framework of different entities which is spread across various activities monitored and controlled by market and legal norms and regulations. All these activities must be duly fulfilled, otherwise, it will lead to collapse or damage to food supply chain. The following sections in this chapter will provide an idea and explanation of sustainability, following which the aim, objective, and methods of controlling the Sustainability of Food Supply Chain will be explained. Furthermore, the need for controlling and its associated benefits will be detailed.

6.2. CONCEPT OF SUSTAINABILITY

In most of the regions in the world, the imported foods and a mix of local production impacts eating habits. Currently, the food manufacturing has become a significant part in the diet of most of the people. Also, most of the staple foods in the world are traded as commodities and products in international markets. The local food supply chains are considered as partly sustainable as they help in supporting organic and mixed farming as well as lower toxic externalities and emissions. Local supply chains of food are valued for developing rural enterprises as well as regenerating rural communities, develop spiritual links and associations and break the monopolies of agribusiness between nature and man (Figure 6.1).

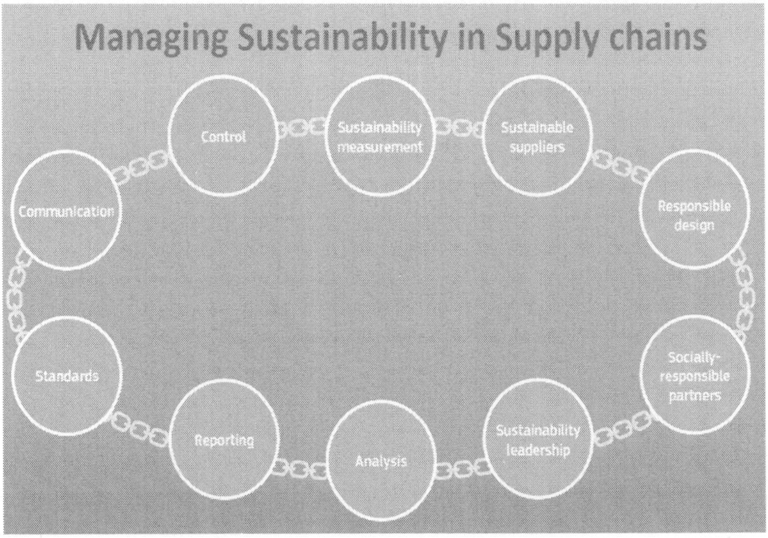

Figure 6.1: Managing sustainability in supply chain.

Source: https://ec.europa.eu/europeaid/sectors/economic-growth/private-sector-development/sustainable-and-responsible-supply-chains_en

Conservation of food is significant to reduce degradation and losses at the time of transport from urban to rural populations. It also allows people to understand the variety and nutrition in a diet all through the year. The food products which are stable are produced by fermenting, smoking, salting, and drying and are stored and retailed in the outer local region in a preserved food supply chain for many years. Freezing, pasteurization, and canning has raised a variety in foods which may be conserved. Also, the latest technologies like controlled atmosphere, aseptic, and chilling packaging have led to conserve most of the "fresh" fruit and vegetables for a long period of time. Also, dairy foods are traded and retailed as conserved food, having a high impact over the international and national patterns for trade and production of agriculture.

Foods which are manufactured consist of ingredients sourced from different production systems and origins. Although few may be derived through simple supply chains, others can include most of the transactions in manufacturers and farmers, it is a chance that aggregate quantity of transactions included in a single product is quite high. Some of the ingredients like dried fruits, spices, and herbs can build only a small portion of finished goods or can be needed for a shorter duration of product life of a specific product variety. Also, a majority of the ingredients can be derived through a changing network of international supply which may overlap and coincide with the competitors of the business. For instance, the vegetable oil used in almost every kitchen is substituted for compensating for the occasional scarcity or difference in price and quantity. Substitution and mixing in a complex dynamic network of supply can restrain traceability and influence the information flow in a chain.

6.2.1. What is Sustainable Food Supply Chain?

According to UK Sustainable Development Commission, there are many different views of stakeholders which are combined for developing a description and explanation of Sustainable Food Supply Chain which has its application in international market as well.

- Produce healthy and safe products as per the demand in market as well as ensuring the easy access for the customers towards nutritious food along with accuracy information regarding food products.
- Support the diversity and viability of urban and rural communities and economies.
- Enable feasible livelihoods produced from managing and sustaining land, through payments and the market for the advantages of public.
- Respectfully execute in the biological boundaries of all the natural resources like biodiversity, water, and soil.
- Attain continuous high standards and limits of performance of performance by lowering the consumption of energy, lowering the inputs of resources and making use of renewable energy.
- Make sure for attaining a hygienic and safe environment to work as well as high training and social welfare for the employees who are a part of food chain.
- Getting a high standard for the welfare, hygiene, and health of animals.
- Maintain and sustain the available resources to grow the food as well as supply benefits of public, only except the situations

wherein alternative uses of land are required for meeting requirements of society.

A variety of actors which are working inside and exterior to the food supply chains which are required to make necessary changes and alternative to the ways where they work for making the process of supply chains sustainable. Food businesses have huge impact on consumers as they choose the foods for promoting and making it available by methods of pricing, product placements, packaging, and advertising (Figure 6.2).

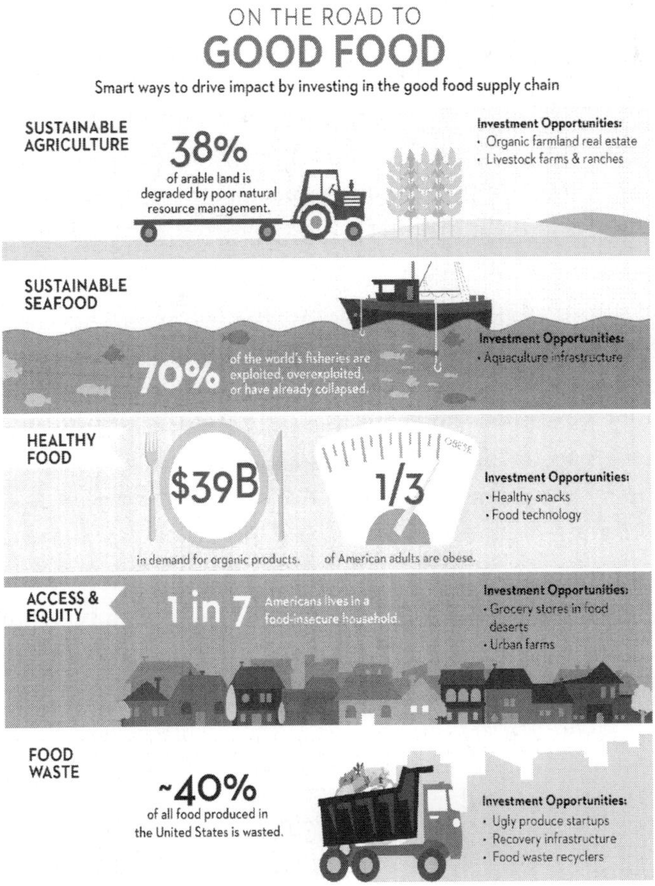

Figure 6.2: On the road to good food.

Source: https://ec.europa.eu/europeaid/sectors/economic-growth/private-sector-development/sustainable-and-responsible-supply-chains_en

6.2.2. Sustainable Food and Beverage Supply Chain

Each day, the increasing population of globe is consuming a huge quantity of food. Managing the food supply chain are in an intense pressure for the purpose of feeding the globe. The present method of increasing most of the food products cannot be considered as sustainable. Worldwide production of food is required to be increased by approx. 60% for meeting the expected demand and need from 9 billion of population. It is necessary to make sure that supply of the raw materials is reaching to all the food producers and manufacturers. The food industry is required to be under monitoring of whole of the supply chain and it should work along with the suppliers for improving the quality and yield as well as lower the harmful impact on environment. The sustainability in supply chain of food and beverage is required which is done by making use of advanced software which helps in delivering the visibility and traceability. When the chain ends, the consumers demand a deeper insight in the root of products which they purchase. They make sure that they buy a product which does not damage the producers or environment.

For the actors of supply chain, either small or large, the main task involves moving towards actions for developing sustainable practices in business having lower influence over the environment besides, producing plenty of food to supply. There is a noticeable scarcity in water and land resources which are of degraded quality in most of the segments of world. As there is climatic change, there will be a lot of challenge to fulfill the needs of increasing population.

Though there are various challenges associated to food chain sustainability, a better and more justifiable food supply chain brings in various advantages which include traceability and visibility. Few of all the actions which are used in the food supply chain management cover broad variety of applications like an innovative software solution for predicting the shelf life of products in food industry, analytics, and residue governance, assessing the suppliers and traceability.

6.2.3. Management of Food Supply Chain and Sustainability

In this industry there is an advantage for the retailers and manufacturers for taking independent actions and be in competition by making claims to have unique, best, and more sustainable and nutritious products. The food supply chain is becoming sustainable every year on the basis of reports which track and follow these initiatives. The food manufacturers, retailers, and distributors advertise their successful achievements and victories in annual reports of sustainability. Since all the players polish their efforts, organizations which facilitate and enable product movement logistics service providers have a significant supporting role especially in the visibility area (Figure 6.3).

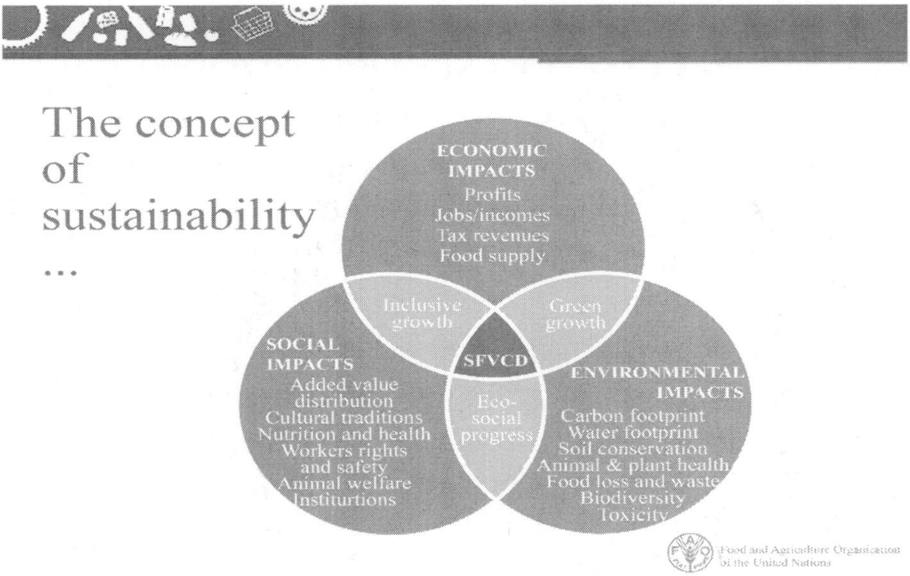

Figure 6.3: Sustainable food value chain.

Source: https://www.slideshare.net/ppanth/14-sustainable-development-concept

Most of the sustainability issues like use of water, agricultural investment, declined space of farm, worker's treatment, organisms which are genetically modified, and deforestation fall over processing and agriculture instead of on food movement. Many observers have argued that Supply chain has a significant role because organizations identify the relevance of visibility of supply chain to improve the sustainability determinations.

According to the Vice President of Analytics and Supply Chain Services at Havi Global Solutions, Pan Chen "Selecting and choosing the raw materials is considered as a core part of function of supply chain. As the actions can be placed in manufacturing and farm facilities and conveniences, it depends on professional of supply chain to assess the location and quantity of source materials. He also explored and determined the opportunities for lowering the waste by using dynamic optimization of supply chain. Some of them included: assessing all favorable and feasible geographic outline of location and whereabouts of supplier, their distribution centers, forward warehouses and manufacturing facilities for maximization of efficiencies.

6.3. KEY CHALLENGES AND PRIORITIES IN CONTROLLING SUSTAINABLE FOOD SUPPLY CHAIN MANAGEMENT

When there is a change in market conditions along with fluctuating demand and available supply, the supply chain can be adapted accordingly by making real time adjustments. The right infrastructure must be a priority for giving the customers transparency according to their need.

6.3.1. Key Challenge: Visibility

The core sustainability challenge involves improving and betterment of supply chain in all its phases. According to the Impact Report in 2016 by Sustainability Consortium, a huge number of manufacturers of consumer goods lack the vision in sustainability performance and actions of supply chain which is the primary barrier for improvising the sustainability. The products related to agriculture and food have a huge upstream share where there is lower visibility. It shows a high potential risk, however there is also an opportunity for improvement in sustainability and costs. As per the consortium data, in grape, apple, and berry categories, approx. 60% of the respondent were able to define that whether the suppliers obtained the crops from lands of excessive carbon stock in previous five years. In opposite to it, less than a quarter of egg suppliers, dairy, beef, chicken, and pork made similar determination and assessment for the supply need (Figure 6.4).

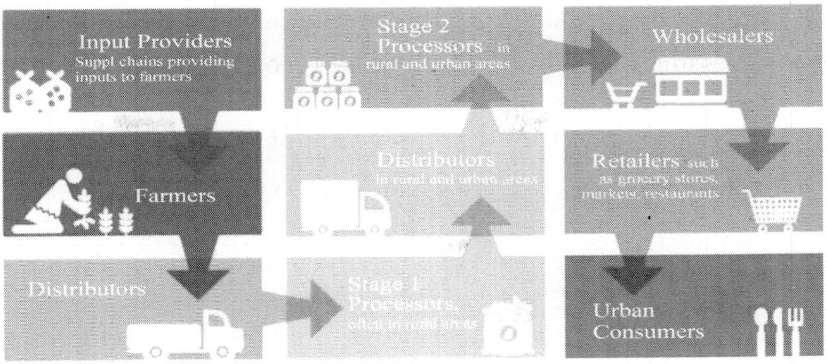

Source: AfDB, 2013

Figure 6.4: Agri-food system supply chain.

Source: https://www.australianfutures.org/s/Ripe-for-Change-Final-Report–160223.pdf

The requirements for improving the visibility of supply chain has caused and led to various initiatives. Packagers, processors, and growers have accepted various parameters to improve sustainability. The community of regulations who respond to concerns of the consumers regarding social justice, safety, nutrition, and environment has developed more challenges.

There are many sustainability challenges which are in headlines like using antibiotics and suppression in supply chain. Lawsuits and regulations can either encourage or need food packagers for including the labels and tags which address and take care of these issues.

Large companies into food and beverages are executing when planet, the economic landscape and consumer expectations experience significant and rapid changes. The company's business models are entwined with most suppressing global challenges faced today like depletion of resources, scarcity in water, food insecurity and poverty of most of which are affected through shifting lifestyles, climate change and rapid population growth.

The long-term investors expect the companies to disclose and manage the governance, social, and environment impacts as proof of efficient governance incorporates. Also, the investors are depicting specific interest in the way the industry of food and beverage is replying to challenges of agricultural supply chain. This is because land disputes, poor working conditions, extreme weather and drought are driving and moving the financially relevant risks in business for the companies related to food like seizure of goods, legal sanctions, harming the brand equity from legal campaigns, supply of ingredients, inconsistent quality and price volatility.

6.4. IMPROVISING SUSTAINABILITY IN FOOD SUPPLY AND AGRICULTURAL CHAINS

The three main essential components of responsible natural resource management, protection of environment and animal welfare are required for continuing producing sufficient food for quickly increasing population in globe in the future. There is a huge contribution from the banking sector as well for bringing more sustainability in food production and agriculture which may benefit the society, environment, and economy.

The population all over the world may increase from 7 billion to 9 billion till 2050. As per this trend, huge population can observe the increase in their income. This may lead to global food production to be increased by minimum 60% for achieving the increasing food demand. Simultaneously, natural resources like food and water and land will also get scarce in comparison to their availability at present (Figure 6.5).

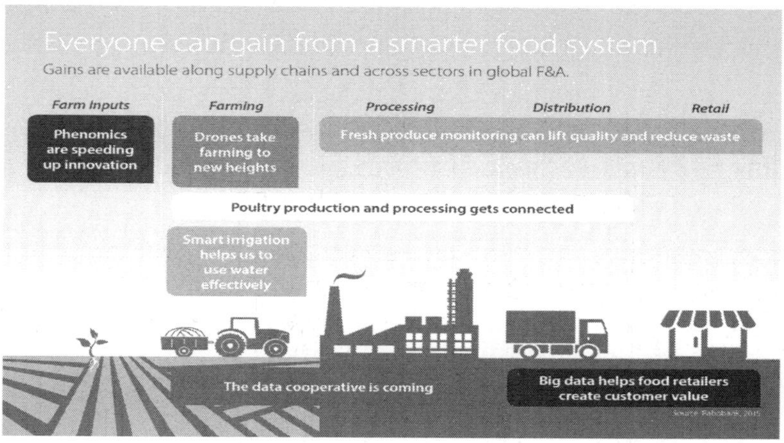

Figure 6.5: Everyone can gain from a smarter food system.

Source: https://www.australianfutures.org/s/Ripe-for-Change-Final-Report–160223.pdf

Food chains and agricultural companies are required for increasing the output, however, the access is lesser towards animal and human wellbeing and natural resources. For the agricultural and food sector, an increasing population and depleting arable land is in need of controlling the agricultural inputs sustainability for the industry. Controlling of sustainability if of food products involves controlling and monitoring the environmental, social, and economic performance and deliverables of supply chain.

6.5. SUSTAINABILITY AND RURAL WORLD

Rural world experiences a huge challenge which is a home to nearly three-fourth of poorest people in the world and also faces issues regarding limited resources like migration of younger people, land, and water. The report involves studies which were conducted in project of "Sustainable Development in 21st Century" which was an undertaking of United Nation Department's Sustainable Development Division for economic and social welfare. The main objective of this project comprised of constructing and developing a cohesive vision for sustainable development of food supply chain in 21st Century. The main questions which are addressed in this section of the chapter is: what are few of the most significant latest trends in agriculture and food since the beginning of Original Summit of Rio in 1992, the actions that have occurred in subsequent 20 years for enduring sustainable systems of agriculture and food and the top issues to be the priority list items.

6.5.1. Trend of Undernourished People

Of the 1.1 billion in world of total poor people, nearly, 74% are residing in the areas which are completely marginalized and also rely over small scaled agriculture. While the globe presently develops and manufactures sufficient food for feeding everyone, minimum one billion of people are left as food insecure. Though the situation of hunger was lowered to one from three to affect out of seven people in 1990s, the trend was opposed in 1990. Sub Saharan Africa has huge proportion and ratio of undernourished people, 60% of undernourished people in world are living in nearly 7 nations: Pakistan, Indonesia, India, Ethiopia, democratic Republic of Congo and Bangladesh Approximately all nations have some degree of food insecurity level (Figure 6.6).

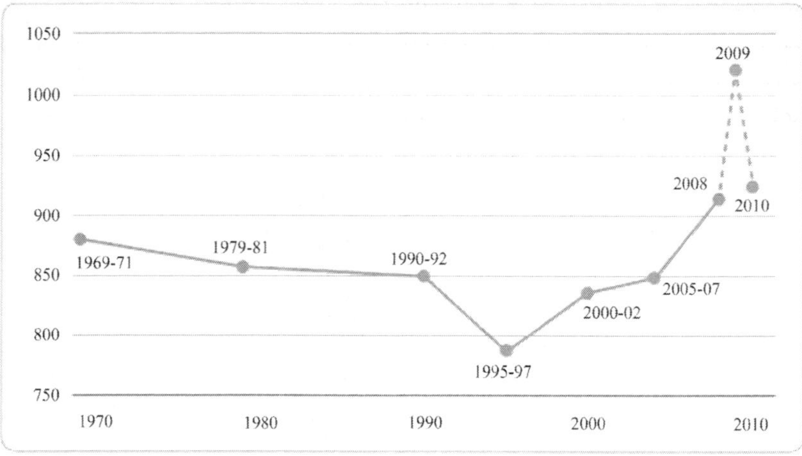

Figure 6.6: *Trend of undernourished people (million) 1970–2010.*

Source: FAO, 2011. Rural Poverty Report. Rome. Using other FAO publications sources: "State of Food Insecurity in the World" and "Global Hunger Declining but still Unacceptable High."

Global agricultural production grew at an average rate of approximately 2% in a year in between 1961 and 2007.

The unpredicted rise of nearly 6 decades was caused mainly from the increase in yield as well as from expanding the land area in cultivation and irrigation. Also, it has been estimated by United Nation that if the present patterns of dietary consumption are consistent over similar trajectory, there is a need to feed more than 9 billion populations by 2050 which means that we are required to generate the food as much as we have in previous 8–10000.

6.6. STRATEGIES AND METHODS TO IMPROVE FOOD SUPPLY CHAIN FOR ENHANCING THE SUSTAINABILITY AND RESILIENCE

Strategies and methods to improve managing the food supply chain for enhancing the sustainability and resilience involve management of activities performed in industry. Land disputes, habitat destruction, ecosystem degradation, insufficient labor protection, dangerous, and unsafe working conditions, human trafficking, forced labor and child labor are some of the undesired impacts and consequences over the industrial activity. Companies in almost all the sectors like food and technical business are in a pressure for being transparent regarding how they will plan for mitigating potential environmental and social risks and influence over the supply chain (Figure 6.7).

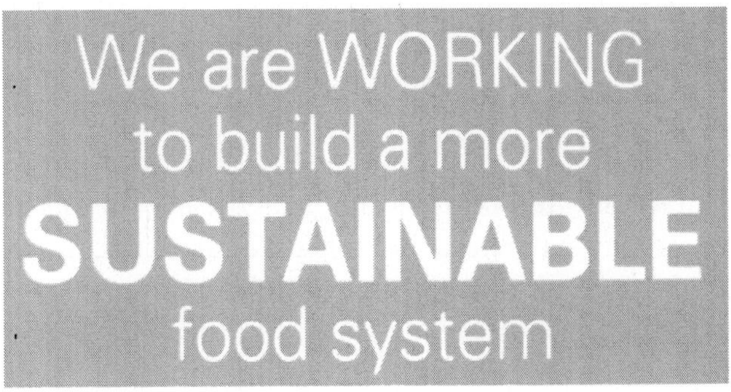

Figure 6.7: Improving sustainability.

Source: http://www.incae.edu/clacds/supplychain/

Having regulations like UK's Modern Slavery Act of 2015 as well as California Transparency Supply Chains at of 2010, the corporations are required to be transparent regarding the influence of activities as well as the efforts for addressing the risks in supply chains. Most of the companies struggle and make efforts to understand and identify their supply chains, which makes them exposed towards criticism that the people are not able to manage the embedded environmental and social risks. The non-government organizations and news media and reports have mentioned about the cases of poor conditions of working, damage to the environment and child and forced labor where the commodities of the agriculture are harvested and grown and where the fish are processed and caught. The sustainable supply chain of food is required to address and cater to the environmental and social issues in the complex structures. Food supply chains are required to address some of the key challenges: complexity, environmental issues and social issues. All these three challenges are described below:

6.6.1. Social Issues

In social issues, the workers are exposed to worst conditions in their labor hood like child labor and forced labor that are closer to the beginning of supply chain. All these conditions are present due to the nations where the food ingredients are processed and have unenforced and inadequate judicial systems, regulations or due to requirement of seasonal unskilled labor by processing and farms facilities. Most of the temporary workers and labors are contracted not through the companies, who understand the process of their hiring, but through brokers. These unskilled and low wage labors are taken benefits of as they do not have knowledge of terms and conditions for their contract as well as are unaware of the rights.

6.6.2. Environmental Issues

Since the plantations and farmers are specifically in rural regions, the practices of them may have negative impact on ecological and biodiversity services. All the negative impacts may be through encroaching the plantations and farms and because of using fertilizers and pesticides. They are identified upstreaming of supply chains which makes them tough to be found. The ocean ecosystems and fish stocks are also adversely impacted in the fisheries.

6.6.3. Complexity of Supply Chain

When the organizations identify suppliers in food value chains, the companies start understanding the complexity and intricacy of multiple supply systems. For instance, the agricultural products are bought from a single farmer and imported through many farmers overseas or even bought from intermediaries like traders and cooperatives. When the inputs that are processed are bought, a unique worksite is developed in the food chain. They are existent at different points in the value chain. Additionally, any food products are not made using a single ingredient to increase the complexity by multiplying it with various ingredients in end product. When a product is seen by various parties and is consolidated with various other ingredients, the organization find it tough for identifying the supplier and assess where they should target the resources and efforts for developing a sustained supply chain (Figure 6.8).

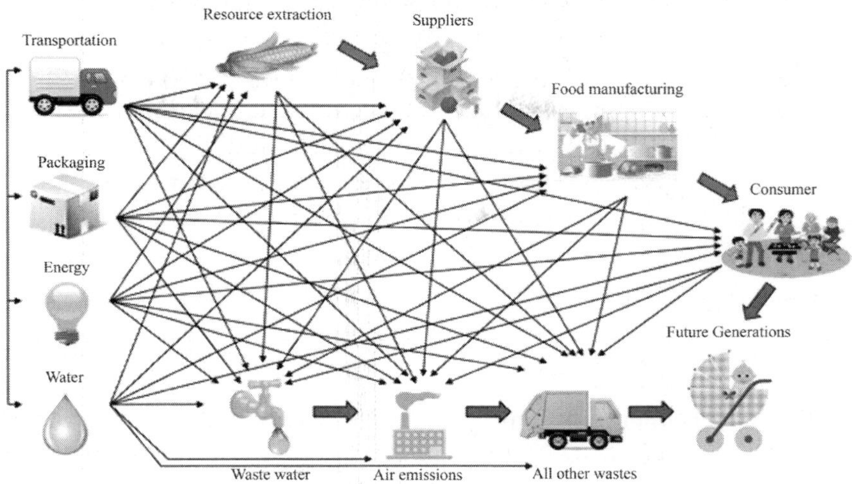

Figure 6.8: Complexity in food supply chain.

Source: https://foodtechconnect.com/2010/07/29/exploring-relationships-in-the-food-system-map/

6.7. EFFECTIVE AND EFFICIENT SUPPLY CHAIN MANAGEMENT

In order for a supply chain to be resilient and sustainable in a company, the supply chain monitoring and controlling should have efficiency in allocating efforts and resources, focused over suppliers who are at risk and must be in control of actions of supply chain and impactful on suppliers.

6.7.1. Be Focused

Companies having high supplier clientele face complexity in managing the supply chains. The company is required to assess how and where to put the resources in various key inputs as well as number of suppliers in multiple levels for maximizing the risk mitigation. The systematic approach and right intelligence may help the company to priorities and focus over the efforts.

For example, most of the countries which produce a huge quantity of common agricultural products like sugar, palm oil and coffee are considered to have bad and unrelenting working conditions (Figure 6.9).

Commodity	No 1 Producer	No. 2 Producer	No. 3 Producer	No 4 Producer	No. 5 Producer
Bananas	India	China	Philippines	Brazil	Ecuador
Dry beans	Myanmar	India	Brazil	Mexico	United States
Cocoa	Ivory Coast	Ghana	Indonesia	Nigeria	Cameroon
Coffee	Brazil	Vietnam	Indonesia	Colombia	India
Palm oil	Indonesia	Malaysia	Thailand	Colombia	Nigeria
Sugarcane	Brazil	India	China	Thailand	Pakistan

Figure 6.9: Production data of common agricultural products.

Source: BSI Screen; FAO STAT, 2013 production data

Through intelligence, a company may identify that dry beans are coming from India or United States. If it is assumed that Brazil is a country with high risk for bad conditions o working, the organization can make decision for subsequent inquiring with suppliers regarding the conditions of the working in the processing plants of dry beans. Also, if the commodities of agriculture are either exported to or imported from nations which are not depicted in Figure 6.9 they still present high social and environmental risks.

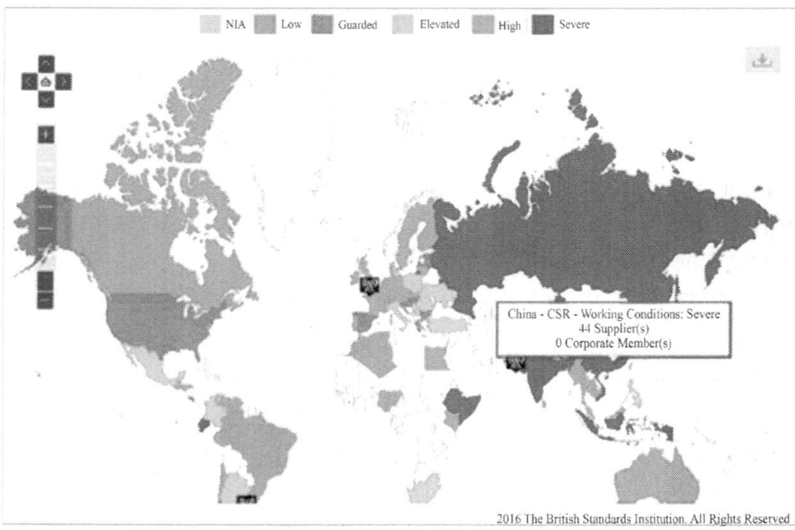

Figure 6.10: Sample supplier management software supply chain management.

Source: https://foodtechconnect.com/2010/07/29/exploring-relationships-in-the-food-system-map/

The software for supplier management, such as supplier compliance manager of BSI, may overlap country-based intelligence like child labor indexes into relevant data which is specific to supplier. When the information is developed in context, the organization may visualize the concentration and location of supplier in its supplier base related to the relevant risks for determining the areas where resources must be concentrate and allocated for optimization of mitigation of risk. For instance, a company may focus and allocate its resources over a huge number of suppliers who are placed in high-risk nations instead of the few suppliers in the nations having a low amount of risk in that country (Figure 6.10).

6.7.2. Be Efficient

If a particular supplier is located in a particular country which has poor working conditions, it may not be assumed that the supplier is struggling with poor conditions of working. In its clarification, an organization can request and demand for a specific information through suppler regarding its practices and actions of labor and associated impact on the society. The companies can send questionnaires for the self-assessment to a population of suppliers or to only a subset of suppliers. The process needs the companies for managing collection of data and flow of information for understanding the risk and also makes good decisions regarding their practices and methods of sources.

The companies can efficiently analyze and gather the information with responsible and well-designed sourcing programs. For instance, the software for supplier compliance is able to store and keep every information at a single place which helps an organization to make comparison in supplier on the basis of primary parameters, make incorporation of own importance criteria and risk tolerance and monitor and track the status of the process of supply chain management. An organization can make efforts in communicating with its reliable suppliers regarding the perquisites for self-assessment and follow ups by using an automated system of email notification (corrective action practices). Verification schedules for audits and onsite visits can also be tracked in the framework and software. All the efficiencies can relieve the sourcing and recruiting team from time-consuming and redundant information communications, analysis, and gathering which allows the team to spend maximum time over working with the critical suppliers who are at high risk.

6.7.3. Be in Control

For increasing the visibility of supply chains, the companies are required to map out or identify suppliers as well as select few for the audits, till all the suppliers are audited easily. When the organizations face a huge supplier base, they choose and select the suppliers only for audits on the basis of available information as well as current supplier relationship. In the given case, the organizations are not under the control of managing supply chain as the decisions and choices are driven through ease of request or even lack of information and not through any imminent risk. By utilization of a systematic way to identify the risks, the companies may adopt two methods: lower the quantity and number of verification audits and onsite visits and therefore the costs which is linked with monitoring activities of supply chain, and also lower the probabilities to overlook the high-risk issues and suppliers. Having a well-designed and a systematic approach for risk-based management of supply chain, the companies and organizations may stay over top of mitigation and risk of supply chain (Figure 6.11).

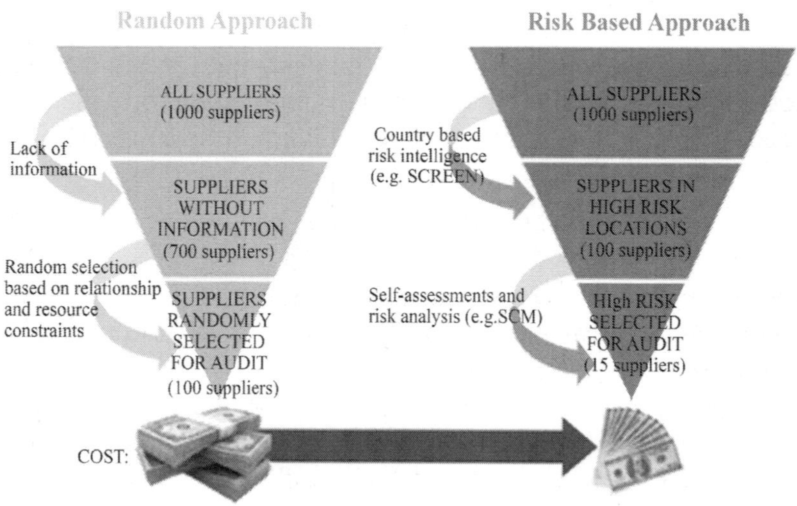

Figure 6.11: Risk-based sustainable supply chain management.

Source: http://fstjournal.org/features/30–3/managing-supply-chains

6.7.4. Be Influential

An automated framework for supply chain allows monitoring,

communication, tracking, analysis, and identification of process of supply chain management. The correct intelligence and tools may motivate the companies for targeting as well as conducting site audits of critical suppliers for improving the visibility of supply chains. It can result in preventive and corrective actions for implementing for the suppliers. It is however not easy for influencing the practices of suppliers.

Implementing the right actions is based on supplier, market conditions and issues along with other variables. It requires identifying and creating partnership and association with other companies which ranges from Non-Governmental Organizations to the competitors for increasing the leverage and improvising the communication skills. Organizations are required to align own intrinsic incentives and promotions for promoting a cohesive and collaborative approach towards the suppliers.

Actions for overcoming the implementation issues cannot be linear from audit outcomes as solutions are required to be tailored. Working with processors, farmers, and external experts for implementation of social and environmental enhancements can be included. Getting in third party gives objective, purposeful, and fresh approaches and techniques. The companies which operate and execute globally tap in resources which are aware about laws, local culture and customs along with an international experience to support companies for better management of supply chain risks.

6.8. HOLISTIC APPROACH

A holistic approach and technique for managing the supply chain risk is required to improve resilience of supply chain. An organization should have clear strategies and policies for assisting its actions:

- Incorporation of strategies for working with suppliers.
- Monitoring its process of supply chain management as well as track the performance of supplier.
- Implementation of systematic approach for analyzing and implementing relevant risk as per its level of tolerance.

By learning about the risks and addressing the areas of high risk as well as preparing ourselves for worst case scenes, the organizations is ready to experience the unforeseen at this time. Even the organization having strong due diligence actions are subject to unforeseen disclosure or incident related to a region or commodity of their sourcing. An organization can react quickly if it is not aware regarding whether the suppliers are linked with a claim, for instance, that slave labor is used in a particular region of Brazil in agricultural sector. The company can take away its own resources from other relevant and profitable work for fighting or protecting itself against public campaigns or allegations and can even move the suppliers away which cause damage and disruption of supply chain (Figure 6.12).

Figure 6.12: Requisites for good food.

Source: https://foodtechconnect.com/2010/07/29/exploring-relationships-in-the-food-system-map/

Technologies are developing at a cutting-edge level. Food Supply Chain Management is recognized by academia and practitioners. Due to Technology Information, dramatic improvement can be seen in Food Supply Chain Management regarding automatic processing of food like pacing, cleaning along with freshness storage. However, this discipline is still not very capable to address and satisfactorily cater to various challenges faced in real life. The main causes for the insufficiencies are due to low levels of executions from farmer, informatory hurdles in different and varied stakeholders and at last inefficient models of decision making. The strategic and informatory decision makers need a cohesive model for increasing aggregate profitability and high revenue while the data which is put into the models are not bothered in most of the models of traditional myopic. For addressing the present challenges and issues, it is required for investigation of better approaches for accommodation of emerging and growing worldwide situations post critically looking at the present Food Supply Chain Management conditions and practices.

With respect to the complications in the food supply chain, few of the significant issues that involved management of holistic lifecycle, greenhouse emission of gas, facility sharing, reusing of resources and waste are still not properly addressed. For instance, in United States of America, approx. 40% of aggregate production of food produced goes as unused which is almost equal to $165 billion. Such wasted food physically impacts environment as the water gets highly polluted. It also raises the emission of CO_2 since a huge level of population will be developed at the time of its deterioration. Therefore, reducing the wastage of food needs actions at all levels of food supply chain such as while producing the food, recycling, selling, storing, and delivering the food items.

6.9. SOLUTION TO CONTROL SUSTAINABILITY OF FOOD SUPPLY CHAIN

With regard to various echelons, the solutions like logistics management systems, warehouse management system, food production management system must be consolidated and interconnected regarding seamless synchronization and sharing of data. Developing techniques can give contribution in system integration in future. Initially, cloud technology is used for integration of segregated sector by making use of lesser resources. Also, it permits stakeholders for accessing various services through software like infrastructure, platform, and software. Using cloud-based solution, collaborative principles of working and information sharing can be attained by making use of internet equipment and basic computing equipment's. Secondly, internet of technology such as smart sensors and Auto-ID are effectively implemented in aerospace and manufacturing industry. Internet of technology-based solutions for Food Supply Chain Management can give whole management of product lifecycle through real time data capturing, tracing the quality and logistics visibility. Also, in an internet of things based environment, all the objects having calculating ability, networking, and sensing ability can make interaction with each other for facilitating decision making and logistics activities in a way which is intelligent, real-time and ubiquitous. Also, big data analytics has got a lot of attention as it is capable of dealing with huge data developed through food supply chain. It can also aid food companies for making the decisions as per the graphs using more accurate and reliable data input through excavation of invaluable knowledge and information that can be used for everyday operations. Using this information, ultimate, and definitive sustainable food supply chain can be achieved through ideal decisions.

6.10. IMPLEMENTATION OF PLAN TO SUSTAIN FOOD SUPPLY CHAIN

For implementing the plan, big organizations play a significant role in renowned food supply chain which moves in sustainable and green direction. For this, collaborating with economically green relationships can cause a win-win situation that huge organizations will obtain economic advantages as well as the members of food supply chain such as subject matter experts can also be given the advantage. The green relationship is on the basis of joint value creation but making use of new business models regarding external and internal green integration that can be enabled through advanced technologies. Therefore, these organizations can take the initial acts for being equipped by IT systems, however down-stream and upstream in the food supply chain may follow up for a green future.

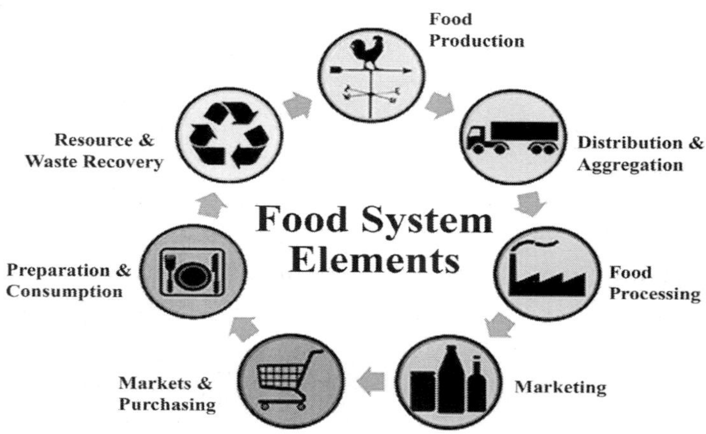

Figure 6.13: Elements of food systems.

Source: http://texaslocalfood.org/food-system/

At last, the implementations require the participation of government bodies that may help in designing the strategic plans to support and guide enterprises for a better future. Therefore, big data analytics is significant to figure out the latest stats report, present status and report of food supply chain and feedbacks from the industry. Also, for identifying the ways and strategies, the companies can make use of advanced models for prediction or even systems for decision main which are data-driven to assist a deep evaluation. Every individual end user can be beneficial from this and its future implementation as well (Figure 6.13).

6.11. ASSOCIATED USE CASE IN SUSTAINABLE FOOD SUPPLY CHAIN

6.11.1. 3S Sustainable Cashew Supply Chain

There is an increasing demand and need for trackable and sustainable food supply chain. Chain point is a company facilitates and helps minimum 100 supply chain in various industrial sectors. Of these projects, one of them helps in improving the traceability of supply chain for cashews and living of cashew farmers in Africa.

Cashew is primarily grown by the smallholders' agriculturists in Brazil, Vietnam, India, and Africa. Presently, Africa produces 40% of the cashew. A bunch of leading participants in the cashew industry has launched and developed a management information system for improving the supply chain of cashew collectively; Securing Sustainable Supply (3S). The 3S Management Information System helps in capturing and measuring the progress as well as impact made over the sustainable objectives and metrics in the supply chain of the cashew industry. The result of this framework will be used for monitoring, evaluating, adjusting, and gearing the improvements in sustainable cashew chain which will be used by local organizations for business development and institutions for training like participating organizations in the farming domain and companies. Additionally, the output will also be used for facilitating the exchange of knowledge and best practices for motivating and encouraging the scaling up and motivating the collective action.

6.11.2. Fresh Point: Traders of Fruit and Vegetables Receive More From the Data Monitoring by Saving Cost Simultaneously

Fresh point is a collective monitoring and controlling platform for residue for international retailers and wholesalers in the European sector of fruit and vegetables. It associates traders and retailers with laboratories and also aids the workflow of assigning, reporting, and analyzing the product tests (Figure 6.14).

Figure 6.14: Supply chain sustainability.

Source: https://foodsecurecanada.org/policy-advocacy/five-big-ideas-better-food-system

The sharing of online functionality and infrastructure as well as sharing and distributing the knowledge over anonymized information and residue monitoring and controlling help the cooperative traders to raise information and process quality, along with lowering the costs simultaneously.

It simply takes a single online form for initiating a time saving and flawless flow of work between food laboratories and traders which examines and evaluates samples of product. The traders in fresh point can also examine sole test results and examine the cohesive reports over total collected data which are generated and produced automatically. They may also use the information for designing effective and efficient sampling plans. Traders can also share the data for understanding the potential risk in their business.

6.11.3. Safety and Quality of Ready to Eat foods

The project of SOPHY aims for developing a web-based software tool to predict shelf life, quality, and product safety of products which are ready to eat. Their decision to choose deli salads and fresh cut salads as their model food system was great. The food producers are stringent in their thoughts for optimization of their selection of raw ingredients, formulation of goods and processing steps effectively. The software helps in predicting the effect of every step of manufacturing over the safety (survival and growth of accurate pathogens) as well as shelf life (which involves the development of particular spoilage organisms as well as the development of unwanted toxic products), by taking into consideration quality parameters. The probabilistic and projecting models and framework are on the basis of current data sets for the change in quality and growth of bacteria in different conditions developed in previous studies and generation of information in a project. The web dominated software has the potential to be expanded consistently-even post the end of the project by way of adding new sets of data like other products of food as well as various conditions of environment. Also, the objective of the project is not only for developing probabilistic models. It involves compilation of information regarding various techniques of processing, formulating the goods, conditions of the environment and managing the safety, quality, wellbeing, and hygiene. The informational sheets have the aims and objectives for educating farmers, as they must not see impact of various factors over growth of pathogen and shelf life calculated by models. This has raised the awareness which is critical to improve aggregate quality and safety of food. It is required for the small sized companies having limited resources (Figure 6.15).

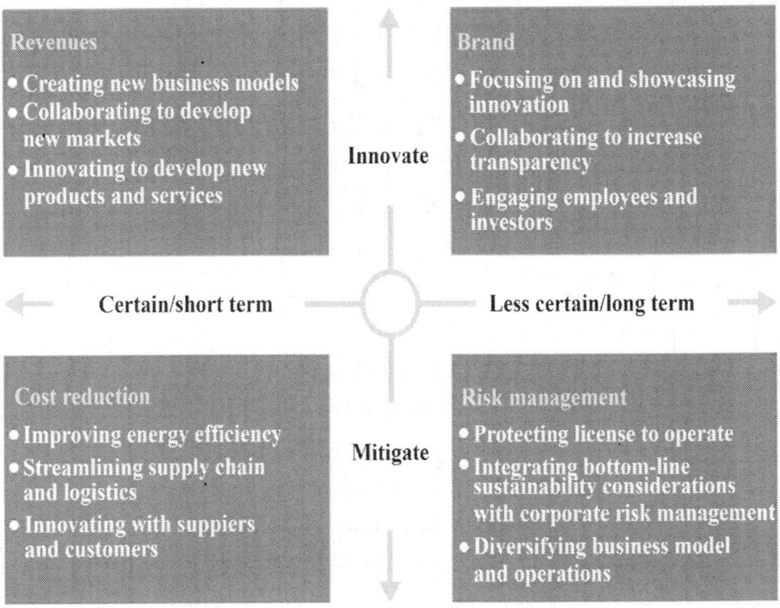

Figure 6.15: Improving food sustainability food supply chain.

Source: https://www.accenture.com/us-en/insight-outlook-why-sustainable-supply-chain-is-good-business

The strategic level of decision making includes all the stakeholders who are keen to participate in a Supply Chain network which is driven through sustainability of agricultural products. Therefore, the decisions made strategically in the span of hierarchy follows different aspects such as: selection of appropriate technologies of farming, Supply Chain partnership association, designing different networks of supply chain, establishing, and forming system of performance management in Agricultural Food Supply Chain and lastly assuring the quality.

6.12. SELECTING THE FARMING TECHNOLOGIES

The present trend for the safety regulations, weather, and biological implications, high concerns for environment, standards of the quality and diversified crops show the requirements for the carefully selecting the technologies of the farming to be deployed. The farming technologies are varied from traditional to sophisticated and contemporary information technology as well as accurate agricultural applications; the latter is known to be a major and significant contributor for growing efficiency in farming and sustainable and environment friendly practices in farming.

The main primary decisions which are included in process of selecting the technologies of farming are linked with:

- Determining the expenditures and requirements of capital on the farming equipment's.
- Developing the cooperative schemes for utilizing farming machinery.
- Adopting innovative and unique applications of farming.

Regarding the cooperative actions and capital expenditure, the accurate solution should be investigated according to the categorization of timeliness costs, labor costs, and operating costs, ownership costs, harvesting methods, tillage practices and planting. With regard to performance and uniqueness and innovation, the elements which impact the choice of technologies in farming may involve volatility of conditions of soil and weather, the desired quality of products of agriculture and quantity of yielded products. The innovative farming technologies helps in ensuring the adequate supply of sufficient goods which may aid a specific food supply chain network to respond to the demand in market in a specific time period. Many studies and researchers have stressed upon the significance of cooperative plan in utilizing the machinery, especially in medium scaled farms, that are categorized by common factors of agriculture like varieties of cultivated crops, labor employability, and environmental impact, type of soil and size of farm.

6.13. CONCLUSION

The issues which are faced by food industry and is at an increasing pace involves expansion of global trade, rising restrictions on agriculture, change in climatic conditions, and an ever-increasing population. In order for an agricultural framework in the world to attain a great balancing practice, the subsequent four decades and periods should match the earlier achievements and attainments in growth of food production without transforming the ecosystems as well as an increasing use of water. Various other elements, from change in climate to the urbanized scarcity of investment, may make it harder and tougher for producing sufficient food. Also, there is a rigid academic consensus which may have serious and long-term effects over the yields of crop like water scarcity which is driven by climatic change, increasing temperature of globe as well as extreme weather. The efficient controlling of sustainable actions in supply chain may not build up over an entirely new parameter of skills. Also, efforts for developing more sustainable actions give the contribution towards consistent development capabilities and competencies of the business. Designing and refining the controls and monitoring of supply chain is a process of continuous improvement which is built upon past as well as new findings and results of the study.

REFERENCES

1. Akkerman, R., Farahani, P., & Grunow, M., (2010). Quality, safety, and sustainability in food distribution: A review of quantitative operations management approaches and challenges. *OR Spectrum*, *32*(4), pp. 863–904.
2. Bianchi, M., (2012). *Balancing Food Safety and Sustainability Opportunities for Co-Management*. [eBook] Available at: http://ucfoodsafety.ucdavis.edu/files/150364.pdf [Accessed 17 Mar. 2018].
3. Heikkurinen, P., Jalkanen, L., Matti, J., & Maaria, U. K., (2011). *Corporate Responsibility in the Food Chain: The Criteria and Indicators*. [online] Available at: http://centmapress.ilb.uni-bonn.de/ojs/index.php/proceedings/article/viewFile/1247/247 [Accessed 14 Mar. 2018].
4. Lee, C., (2017). *What is In Vitro Meat?* [online] Office for science and society. Available at: https://www.mcgill.ca/oss/article/technology-health-and-nutrition/what-vitro-meat [Accessed 16 Mar. 2018].
5. Lewin, K., (2017). *Food Trends 2017: How Technology is Shaping Our Future*. [online] Creativepool.com. Available at: https://creativepool.com/magazine/technology/food-trends–2017-how-technology-is-shaping-our-future.11960 [Accessed 16 Mar. 2018].
6. Maloni, M., & Brown, M., (2006). *Corporate Social Responsibility in the Supply Chain | Food Communication*. [online] Sites.jmu.edu. Available at: http://sites.jmu.edu/foodcomm/bibliography/maloni-and-brown-corporate-social-responsibility-in-the-supply-chain/ [Accessed 14 Mar. 2018].
7. *Shaping the Future of Global Food Systems: A Scenarios Analysis*. (2017). [eBook] Available at: http://www3.weforum.org/docs/IP/2016/NVA/WEF_FSA_FutureofGlobalFoodSystems.pdf [Accessed 16 Mar. 2018].

CHAPTER 7

CORPORATE SOCIAL RESPONSIBILITY IN THE SUPPLY CHAIN: AN APPLICATION IN THE FOOD INDUSTRY

"It seems to me that our three basic needs, for food and security and love, are so mixed and mingled entwined that we cannot straightly think of one without the other."

—M. F. K. Fisher

CONTENTS

The food industry has been a part of all those sectors which has shown accountability towards environmental and social sustainability. It is important for the food industry to understand that how their organization can become more ecologically sustainable, economically competitive and socially responsible. In this chapter below, a detailed discussion has been done on all the social aspects of the food supply chain. Corporate Social Responsibilities is that mechanism which would help this industry gain more public visibility and would also help in getting connected with the consumers. In this chapter below, a section has been dedicated towards the importance of CSR in food supply chain, as this industry is fully aware of its social responsibility and is committed to making a positive contribution to society and the natural environment.

7.1. INTRODUCTION

In Business Studies curriculum terms, Corporate Social Responsibility (CSR) can be defined as the activities undertaken by the businesses which are looked upon beyond profits for shareholders or the owners. A stakeholder is somebody who plays an important role in taking up the final decision for a business. Corporate social responsibility (CSR) has now become an important issue in the food industry. Michael Maloni and Michael E. Brown conducted a research on Corporate Social Responsibility in the Supply Chain: An Application in the Food Industry and stated in their journal that "A multitude of stakeholders in the food sector are increasingly interested in corporate responsibility (CR) and business impacts on society and natural environmental" (Maloni and Brown, 2006).

The major challenge currently in the supply chain of food sector is Sustainability which affects the food industry in a different manner and tackling these issue in context to Corporate Social Responsibility (CSR) has been an increasingly popular practice amongst the food industry. Corporate Social Responsibility are those activities which help in maintaining a balance between a business and its stakeholders in a proactive, professional, responsible, and ultimately sustainable way.

Conducting CSR activities for any organization is not an easy task to perform and such issues are applicable to various types of businesses. But, CSR activities are essential for the food industry as any decision related to food production is going to determine the fate of many species. For example, many of the CSR challenges facing the food industry are reflected in the chicken supply chain. In order to have an effective CSR strategy for a business it is important to follow a few rules, which make the volunteers act for the business and should not just focus upon the selective public relations. The focus should be on the relevant issues for the businesses. The CSR agenda can produce many benefits to firms in terms of competitive advantage, risk management, avoidance of cost and knock-on benefits.

Maloni and Brown also stated that "it is important for the company to exhibit good corporate social responsibility to make sure that their product is made in a way that doesn't harm people, animals (treat them humanely), and the environment." Such activities are not financially beneficial for any organization (food sector) indulged in conducting CSR activities, but they would be worth it both for others and for themselves if word ever leaked about their negative actions. Companies would save more money if they naturally exhibit corporate social responsibility than if they were caught doing something morally wrong and their product was boycotted.

For companies like Nestle, who has its suppliers from various countries (comprising developing countries, also poorer regions of the world), it is essential for them to take a broader view of responsibilities. It is visible that Nestlé is performing successfully from quite a long time and has created value for society. This creation of value for society can be termed as "Creating Shared Value." It aims at creating a new and greater value for society and stakeholders. It further helps in rendering a great deal of impact at nutrition, water, and rural development. According to Maignan et al., 2002, companies go for CSR for various reasons. Based on organizational values, some business leaders have embraced the concept and seek to provide leadership in the area.

7.2. CORPORATE SOCIAL RESPONSIBILITY AND ITS PRINCIPLES

Carroll was the first researcher to describe the classical model of corporate social responsibility which comprises core components related to it. These components have been defined within a pyramid which has been elaborated in Figure 7.1.

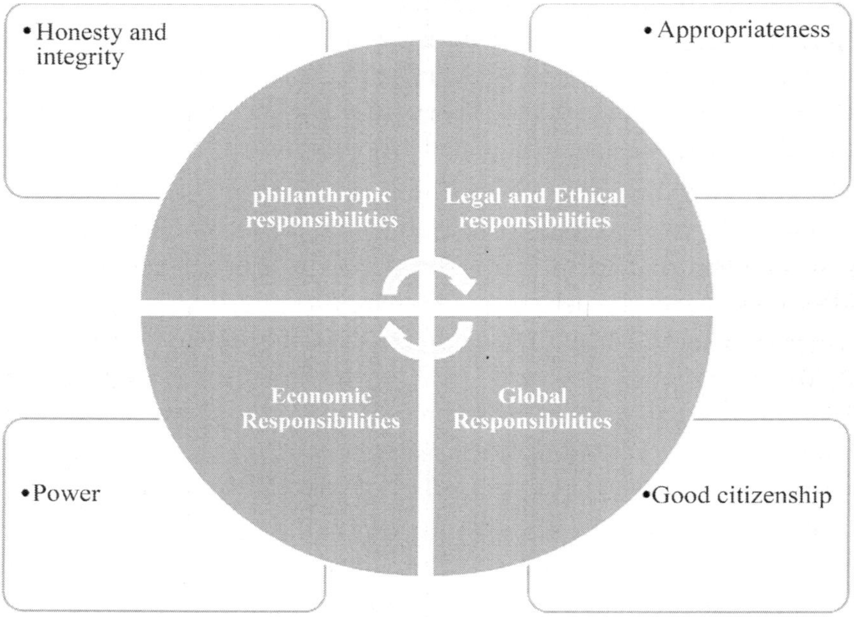

Figure 7.1: Drivers of the CSR pyramid model for a food industry.

Source: https://media.springernature.com/lw785/springer-static/image/art%3A10.1186%2Fs40991–017–0015- y/MediaObjects/40991_2017_15_Fig8_HTML.gif

Carroll focused on four basic points. Almost all the food companies which are socially responsible complies with the laws and carries out the ethical responsibilities which further contributes towards the betterment of the society through philanthropic activities conducted by them. Starting from the bottom of the pyramid which holds the highest level of importance while conducting any CSR activity is the economic responsibility. The highest priority is given to the economic responsibilities led by these activities for sustaining the food chain or the supply chain. The economic responsibilities act as the foundation for the CSR. It is pretty much explanatory that no food company could survive without any profitability. Therefore, its main agenda is to conduct commercial activities in a free enterprise system.

Carroll in his report stated that "good corporate citizens earn enough money that their investors receive a strong return on their investments, and that other stakeholder are assured of the continuity of the business and the flow of products, services, jobs, and other benefits provided by the company."

All the products provided by any food industry, which are profitable to its business must contribute towards the economic growth of the industry. Since a very long time, it has been argued that, apart from supplying products and services to the potential customers for economic growth, the social values are equally important while conducting operations. Carroll also believes that "public expectation has been the primary driver for CSR existence. In brief, public expectations have broadened and so have businesses responsibilities." Besides all this, there are possibilities businesses have to sometimes, renounce its profit for the sake of social interest.

Another important component of this pyramid is the legal component of CSR. To be responsible and also to have its creditability in the market and social environment, it is important for the company to follow basic laws as implemented by the government. The state can contribute towards the implementation of CSR activities vis-à-vis laws. In context to universal laws, code of conduct or compendiums (UN Global Compact) and self-regulation must be followed. But there can be situations when a company would not be able to go ahead with the law compliance. This leads towards the third component of the pyramid which can be described as the ethical responsibility.

Ethical responsibilities come with various aspects of CSR activities from the social component, i.e., reasonable treatment towards the stakeholders to the environmental one with respect to the environment. Business ethics are somewhere related to the bifurcation between fair corporate behavior and unfair behavior. It is expected from the food industry that they must demonstrate integrity and must 'do the right thing'; as it is seen as a moral act. The main strategy of businesses for including ethics in CSR is to adopt code of conduct within their company and bring transparency by releasing sustainable reports.

The top-most component of Carroll's pyramid has been dedicated towards the philanthropic responsibility. As stated in the journal, Corporate Social Responsibility in food industry, good corporate citizens give back to the communities in which they reside or maintain offices. At the corporate level, philanthropy can be done by proving funds or by volunteering.

The first two components of Carroll's pyramid are required to meet the legal aspect of the business and also CSR undertaken by them. CSR is more about assessing and understanding businesses' impact on the environment as well as its effect on social welfare (i.e., elaborate a complete risk management), secondly, reaching a right balance between the social, economic, legal, and environmental responsibilities a company has; and thirdly, it is the integration of all level of these responsibilities with the addition of ethical value into the business strategy before a business can call itself socially responsible.

7.3. FOOD CHAIN IN CONTEXT TO CSR

Melanie Fritz and Gerhard Schiefer, in an article named Sustainability in Food Networks, 2009, stated that food companies work and operate in a very complex and multidimensional context which can highly impact the supply chain in the long run. Many researchers claim that the food industry is going through its revolutionary phase. The food chain comprises primarily the producers who go through a difficult phase and challenges when compared with the manufacturers and retailers. These producers are placed at the top when looked upon the food chain and manufacturers and retailers are positioned further down the chain.

The agricultural food system has a boundary which cannot be crossed easily by the private companies. The scarcity of natural resources creates different kinds of hurdles for food companies. Those hurdles are in the form of price hike or natural resource scarcity, for example, the moment there is a shortage of any natural resource in the market, its prices get automatically hiked exploiting the middle-class and poor people, such as shortage of oil and phosphate, are likely to skyrocket.

This illustrates that all food companies, chains, and networks must change their approach towards the society and should bring more responsible business practices that contributes to sustainable development. The quest for defining corporate social responsibility in the food chain is founded on an attempt to measure the performance of these responsibilities held by the companies. This measurement can provide with an idea of what is exactly required in terms of resources, i.e., natural, human, and environmental capital, and what kind of progress should be made and has already taken place. It would also help in measuring the outputs of the process and in what manner these CSRs have helped the companies (in terms of environment, society, and profitability). In addition, performance measures can be used as the basis for communicating issues related to responsibility. For these reasons, work on measuring and defining responsibility has great relevance for business leaders and managers but also for other stakeholders, such as state representatives and non-governmental organizations (NGOs) (Figure 7.2).

		Supply chain			
		Primary production	Manufacturing	Retail	Consumer
Corporate responsibility (CR)	Environment	Measuring responsibility in terms of inputs, process, outputs and impacts			Demand for responsibility information
	Product safety				
	Nutrition				
	Occupational welfare				
	Animal welfare				
	Local well-being				
	Economy				
		Other stakeholders			

Figure 7.2: Conceptual frame for corporate social responsibility in the food chain.

Source: http://centmapress.ilb.uni-bonn.de/ojs/index.php/proceedings/article/viewFile/1247/247

The conceptual frame for corporate social responsibility has been described in the above Figure 7.2. The supply chain has been simplified in to four different sectors primarily known as, primary production, manufacturing, retail, and consumer. The relevant stakeholders are depicted at the bottom of the food chain comprising governmental actors such as non-governmental actors such as activist groups, as well as consumer agency. There are seven dimensions to the food chain in respect to the consumer responsibility an organization holds. Those seven dimensions are: environment, product safety, nutrition, occupational welfare, animal welfare, local well-being and economy. For measuring the responsibility undertaken by any organization for the supply chain under food industry in followed by the four principles of measuring. Those four principles are: inputs, process, outputs, and outcomes. The consumers are the ones who demand for all the information related to these responsibilities carried by the food sector.

The food industry makes efforts through CSR to retain substantial public visibility since it not only supports a requirement of daily human life but also plays a large role in the national economy as a multi-trillion-dollar industry and leading export. With a complex, multi-echelon supply chain, the US food industry has many distinct and problematic supply chain issues related to corporate social responsibilities that remain in various stages of development.

7.4. SOCIAL ASPECTS OF THE SUPPLY CHAIN: AN APPLICATION IN THE FOOD INDUSTRY

The application of CSR in the food supply chain has emerged from the past 15 years. The concept of social responsibilities in context to CSR has gained momentum and companies are making efforts to understand different social aspects of the food supply chain to get connected with the consumers from all around the world and especially those who are located in remote areas. People living in the Northern regions of America enjoys a higher standard of living but suffers a lot in terms of overnutrition and obesity. Around one-fourth of the population is surviving on less than $1 per day and have no solution to the malnutrition dilemma, where the world is full of food and grains. "Approximately 18,000 children die each day from hunger. Projections are that 1.2 billion people may be hungry by the year 2025." But this extensive malnutrition, is only one component of the health and safety concerns with the current food system.

The use of growth-promoting antibiotics in animal agriculture is one of the leading factors driving the increase in antibiotic resistance in humans. Anti-biotic/anti-microbial resistance is considered as a major public health crisis, which would help in wearing down the discovery of antibiotics/ antimicrobials and their application to clinical medicine. But, the use of antibiotics in livestock has various added advantages to it/ According to Mathew et al, 2007, "it is equally clear that the use of the sesame antibiotics leads to the increase of antibiotic-resistant bacteria of human significance." The use of agricultural anti-biotic represents a major driver of antibiotic resistance. This is due to four different reasons:

(1) The first factor responsible for it is due the largest use of anti-biotic at a wider scope in global market.

(2) A lot use to antibiotics in agricultural sector leads to bring a sub-therapeutic exposure.

(3) Mostly drugs with important and useful clinical class are used in agricultural sector.

(4) Human populations are exposed to antibiotic-resistant pathogens via consumption of animal products as well as through widespread release into the environment.

Wright in 2007 argued that "The first reason related to the evolution and spread of microbial resistance is based on fundamental biology and evolution: the inevitable increase in antimicrobial resistance in response to exposure to antimicrobial agents. Over the millennia, microbes evolved highly effective mechanisms to respond to external agents that threaten their lives." Exposure of bacteria to a level which is incurable for anti-microbial agents is generally effective. The effect that permits the survival of resistant strains and, with continued antimicrobial pressure. All the resistant strains come to have an edge in terms of reproduction and spread. The rapidity of bacterial reproduction and replication allows for greater efficiency in the expression of these changes.

The agricultural sector is in great need for CSR from companies which could provide them with the antibiotics to increase the resistance in the human body. Anything which is ethical and socially as well as economically sound for the agricultural sector would help the sector grow rapidly.

7.4.1. Health and Nutrition

Heller and Keoleian (2000) suggested, "A sustainable food system must be founded on a sustainable diet. In the most general sense, this would be a diet that matched energy intake with energy expenditure while supplying necessary nutrients for a healthy lifestyle." In 2000, the United States food supply provided a statistic which displayed that there is an intake of almost 3,800 calories per person in a day. Accounting for waste, the average person consumed 2,700 calories per day which was estimated to be an increase of 24.5% from 1970. This results in over 154 and 127% more caloric intake than recommended for men and women, respectively. In 2004, it was seen that almost 65% of adults in the United States were either overweight or were suffering with issues related to obesity, defined as having a body mass index of 25 or more, as stated by the University of Michigan, 2006. Irregular diet contributes maximum to heart disease, strokes, and varied kind of cancers, and somewhere these three are the major causes for the death occurring in the United States.

The main cause behind all the chronic diseases is the higher level of saturated fat and low fiber. Almost 67% of average people in the United States have protein from animal sources which is around 34% average intake at a global level. In the 1990s, China reported almost double the number of cases registered for chronic disease due to an increase in the consumption of meat. These chronic diseases were the main cause for the maximum number of death during that period of time. The University of Michigan, 2006, published a report stating that "people in the United States ate 200 lb of meat per person, which is up 22 lb from 1970."

A research by varied public health experts was conducted in 2008 which suggested that "a reduction of around 60% in daily intake of meat in developed countries would help reduce excess weight and obesity and offer other health benefits to individuals and society." This effect will also assist the environmental concerns associated with this type of agricultural production, given that animals inefficiently convert and use energy. For example, the most inefficient animal for energy conversion are the cattle. They require 7 kg of grain to produce at least 1 kg of beef (which when compared to 4: 1 for pork and 2: 1 for chicken) (Brown et al., 1998). Moreover, half of the grains grown are fed to animals.

It is due to three main factors which make us eat food and those three factors are: Conservation of food, its preservation techniques and food processing. People are now highly dependent on preserved food, convenience food or processed food. But, such types of food habits are the main reason behind all the health problems faced by many, as many important nutrients get removed in the whole food products. For example, while producing flour, niacin, folate, iron, thiamine, and riboflavin, are replaced. These processed or preserved food also tends to be high in fat, sugar, and salt since this adds stability and decreases cost. In addition, there seems to be greater enjoyment of the taste and mouth feel of foods high in salt, sugar, and fat, to add to our tendency to consume food.

Companies need to understand all the problems related to health and nutrition of consumers, so that CSR activities are conducted in the lieu of all these components. This would create social awareness and would also have a philanthropic approach towards their consumers.

7.4.1. Small Farms and Rural Communities

Till 1990s, the share of agricultural economy which was provided to the farmers declined from 41% to 9% in the United States. On the other hand, the economic share of the marketing and farm input industries increased by approximately the similar amounts. This reduction in profit for farmers has paved ways for the resources which can also be termed as off-farm resources like, pesticides, fertilizers, hybrid seeds and herbicides. A tremendous increase in the purchase of such resources has been addressed with greater government intervention with respect to price and subsidy. A shift to this kind of trend at a large-scale is also known as monochrome farming. This shift has established too many challenges for the economic health of rural communities bringing in less jobs, less local retail spending, and lower local per capita income for the farmers.

Apart from monochrome farming, another drawback encountered in rural areas is the loss of local food systems. The loss of local food system has also seen to have a different dimension, since direct links between people and the farms that produce their food can enable people to reengage with the natural world, the local landscape, their local community, and the pleasures of eating fresh food in season. Small farms generally are more diversified than large, industrial farms, providing for more nutritionally dense foods and more environmentally preferable production methods. For example, it has been estimated that the small farms in the United States are almost three times the trees per acre when compared with the larger farms. These small farms contribute around twice as much of their land to cover crops and green manures as mentioned by Rosset, 1999. It has also been recognized that due to the diversification and land stewardship of small farms, the maximum number of produce is agricultural output per acre than larger farms. So, if more attention is paid towards these small farms, then there are possibilities that there will be a wide range of social as well as environmental improvement for the small farms. Basically, the shift required in the market is characterized by quality, and not by quantity. This will further involve higher level of diversification in purchasing of agricultural products from industries which process or preserve food and also the consumers. This will lead towards having a community-oriented approach for rural areas and small farms and this will also result in moving the supply chain from a global level to the local level and would reach towards the remote areas of a region.

The United States has witnessed this shift as more farmer markets and community organizations are marketing more directly to consumers. On the other hand, government would still play a vital role in marketing the food products, also, ecolabels have been playing a much more significant role since they have the power to signal price and quality signals, and their subsequent opportunities and costs (e.g., Rainforest Alliance, Fairtrade). These standards and regulating abilities are expected to grow to meet public demand for quality and transparency.

7.4.2. Consumer Insights and Green Marketing

Each consumer segment has their interest more inclined towards the responsible products. Responsible products are those which are socially and environmentally sound products. Market research conducted on such consumers illustrated that the most loyal set of socially conscious purchasers or consumers is about 10% of the population at an international level. Such consumers or purchasers are well-educated as well as informed group is well informed about the product and are driven by values. However, the maximum number of consumers or purchasers comprise a more reasonable perspective which further make them aim to balance their perspective towards convenience, cost, and making a difference. For this, and the very motivated group of purchasers, trust in the company and the claims or marketing is critical.

Due to all these criteria a "greenwashing" of benefit is uncovered. Greenwashing is more like overpromising or misrepresenting benefits of the product. There are consumers who need motivation to purchase products which are socially responsible. However, they will make such purchases if they meet other values (quality, price, and convenience), and they can see how their contribution helps make a difference.

Generally, consumers want to get awarded for their commitment towards the product they purchase (socially responsible products). As a result, the most effective sustainability-related messages are personalized and credible. Typically, any sort of environmental claim on products follow a generic process which comprises an evolution, beginning with basic product characteristics such as recyclable package, then to differentiated performance such as life cycle–based claims, and at last claims to reflect how responsible and committed the company is, such as a claim about renewable energy. All these are the important part of CSR activities processed by any food industry.

While the last means of communication is a more general message, which would also include various environmental claim which are supposed to be practiced by the companies with complete responsibility. These practices would bring creditability to the company and would help them gain success in the long-run. Such a kind of commitment made by any food industry in its supply chain would help in building trust from consumers' end and it would render rewards and consumers in-return reward such companies. Some of the desired company practices are largely focused on producing energy-efficient products, products with health and safety benefits, support for fair labor and trade, and commitment to environmentally responsible practices.

In the United States, the Federal Trade Commission (FTC) enforces the FTC Act which ensures that adopting exceptionally good marketing strategies for any food industry is neither an unfair practice nor it is misleading, if it is looked upon as a reasonable consumers' point of view. Therefore, it is important to understand the consumers and their preferences as there can be various confusions related to sustainability. A consumer may assume all plastic packaging is recyclable, though only limited types are actually accepted for recycling. The FTC has developed Green Marketing Guides to address typical sustainability-related claims. In general, their guidance is that the message be appropriate, accurate, verified, and informative. Some key questions to consider when developing a message through CSR activities by any food company are as follows:

- What is expressed in the activity they practice and how it can be implied?
- What do we need to substantiate for the claim?
- Is there any requirement for the qualification of the claim?

Answers to these questions would help companies adopt better approach towards the consumer and they would be in a better position to act accordingly.

7.4.3. Interests in Organic Food

One of the fastest growing market segment within the food industry is the segment related to organic foods with sales growing at an annual rate of 21% in 2006 in the United States. Most of the consumers located in North America and Europe demands for organic products. These two regions are considered to be the highest revenue generating regions for the food industry at an international level (around 97% of global revenues). Other regions like Latin America, Australia, and Asia are considered to be the important producers and exporters of organic foods. Though, their internal markets are not yet developed and are going through a developing phase. In 2006, it was viewed that there was a surge in organic food and drink sales in North America, which resulted with the United States retail sales reaching US $16.7 billion in 2006 and projected to grow to US $23.6 billion in 2008.

The United States has the largest market for organic products in the world. Significant level of increase in organic farmland have also made it a top-notch producer for organic products; almost all types of organic crops are grown in the United States. However, imports are necessary because of demand exceeding supply. Latin American countries like Brazil, Mexico, and Argentina export significant quantities of organic foods to the United States market. It is quite evident that organic foods are now present in almost every food retail channel, whether it's a restaurant or it's a mass merchandiser offering hundreds of stock-keeping units of organic products.

Due to this, many consumers from the United States regardless of their income, have started having access to organic foods because these foods are now easily available in the same stores where they shopped processed or preserved food. According to the statistics presented by Organic Trade Association, 2007, "The mass market grocery stores represent the largest single channel for organic products, accounting for 38% of the United States organic food sales in 2006. In addition, another 8% of organic food and beverage sales were through mass merchandisers and club stores. Sales have also increased in food service outlets as well as via internet and mail order sales."

It was also estimated that, in 2006, the fastest growing food category in United States was organic meat sales with around 29% growth, also, a growth in dairy products was also witnessed which was around 25%. Apart from these two segments a tremendous increase in the sale of fruits and vegetable (24% growth) and in the sales of bread and grain (23% growth) was also seen in 2006. 2006 was marked as a year for the highest sales in not just organic food segment but also in non-food categories. An increase in pet food sales (37% growth), household product and cleaner sales (31.6%), fiber sales (27% growth), and supplement sales (26% growth) was perceived, as reported by Organic Trade Association, 2007. The introduction of Canadian national regulations for organic agriculture and organic foods is expected to give the market a boost. The U.S. market has been reporting high growth since the United States Department of Agriculture (USDA) which first applied the) National Organic Program (NOP) in 2002, which ensures the development of national standards for the production of organic products.

Another major driver of market growth in both countries is increasing distribution in conventional grocery channels; large retailers like Wal-Mart, Safeway, and Lob laws are expanding their organic product ranges, with many introducing private-label organic products. The success of the private label of Safeway's organic has contributed in initiating the export of these organic products in Asian Market.

Safeway's Organics' is already a leading brand of organic foods in North America; it is poised to become a global organic brand. Within the United States, some major cities have a higher proportion of consumers who report using organic products than others.

Scarborough Research (2007) found that 35% of San Francisco adults bought organic foods during September 2007. The same month when Scarborough Research was published, an increase in the consumption of organic food was estimated at the growth rate of 32% in Seattle which made it the second leading American city. Other leading cities for the growth in the consumption of organic foods are Portland (27%), Washington, DC, and

Denver (26% each), and San Diego (24%). In the United States overall, almost 17% of all adults are considered to be core organic food consumers.

Scarborough Research, 2007 also found that average income spent on organic food by the consumers is around US $127 on their weekly household grocery bill, which is almost 10% higher than the national average of US $115. "The organic food consumer has a relatively high disposable income, with annual household income of US $86,000 a year, 22% higher than the national average." Scarborough Research also revealed that organic food consumers tend to be young with families. They are 19% more likely than the national average to be aged between 18 and 34, and 13% more likely to have two or more children at home.

Some of the leading social aspects of sustainability include, the nourishment food provides to all the consumers without making them compromise with their health, and also, they should have the ability for diverse communities and farms to prosper. Association of various farms has resulted in bringing concentration of wealth, with the reductions in population, and loss of infrastructure in rural communities, which has further led to greater environmental degradation. One important reasons behind providing nourishment at a global level is for the countries like United States and other industrialized nations in order to bring down the percentage of meat consumption. This would help countries in providing various public health benefits, especially a reduction in heart disease, cancer, other chronic diseases, and further increases in pathogenic illnesses.

Such CSR activities which would take social aspects of the food supply chain into consideration will help in providing more nutritionally dense foods and will further improve health concerns associated with pesticide and herbicide use, which are generally required for monoculture feed grain production. To develop a truly sustainable agriculture and food system, it is essential that people who are socially active must aim at providing at least the minimum nutritional requirements needed by everyone at an international platform. The United Nations from 1948, in its Article 25, identified that food is the basic human right which should be specified with sanitation, clean water, freedom of assembly and movement, and many others. No solution short of this will provide a long-term, sustainable global society in which any individual would want to live. Providing nutrition and best kind of food is a big challenge for the present as well as next generation.

7.5. IMPORTANCE OF CSR IN FOOD SUPPLY CHAIN

Currently, food industries are not adopting the 100-year old model for cheap labor and land. They are now more interested in other doctrines due to which the world would run and grow. Corporate social responsibility is an amalgamation of various activities which would help to steward and bolster the community. Corporate Social duties are not a burden or obligation on any organization or industry, but it is regarded as an important factor responsible in building a better world. Gratefully, the food industries of many countries have taken up the responsibility and that is why this industry is the most respected one in making its CSR efforts. There are many examples in United States for food industry which have taken up this responsibility really well. For example, American biscuit manufacturing companies like Kraft Foods have vision statements of being responsible for future generations. The industry that feeds the world has a significant social mandate on its shoulders, in the new sharing economy we live in. Some of the reasons which makes CSR an important thing for food industry has been discussed below:

7.5.1. Inclusive Growth for the Industry

Growth of an industry at present in not only about profits but is also dependent on different types of customer satisfaction, sustainability, and on the basis of net promoter scores. Companies are counting the cost of invading the future of our world at the expense of exploiting their human capital. Since food manufacturing is a highly capital and labor industry, CSR initiatives that take on multiple forms have been identified as the critical players in the community. Also, social media and rising accountability from consumers and their alertness towards the food products is a threat for many companies, as they will have to be compliant to corporate social responsibilities which have been mandated by governments. There proactiveness towards the betterment of the society should be the main agenda while conducting a CSR activity. Snack food manufacturers in India and across the world like Kellogg's have now earned consumers' trust as their central message while promoting their product in the global market.

If any company take inclusive growth as its foundation, they would earn and utilize all their resources around their facilities and will also conduct events and drives which would bring awareness amongst the consumers regarding the effect junk foods have on human health. This will also spread an awareness that how right eating habits is the pathway for living the right kind of life. Corporate social responsibilities illustrate that how companies live and breathe the canons of humanity. The food processing sector in India contributes around 12% to India's GDP and employs more than 1.6 million workers. If the food industry sets the benchmark in building communities around its operations, other sectors will have to sit up and will have to compete with the great work being done by this industry.

7.5.2. An Initiative towards the Social and Sustainable Change Inspired by the Corporate Vision

The current state of the world is very competitive and challenging which comes with many disruptors springing up by the dime, which threatens many businesses to go in losses or even upend the established businesses. But not always these disruptors work in a negative manner for the businesses, it also affects them in an adverse manner when any initiative has been taken to achieve goal in an unscrupulous manner. Regardless of the company's path to profitability and positive balance sheets, the vision of any company is something which would help in stimulating the long-term, strategic planning that goes above and beyond near-term goals like profit and growth.

But this cannot be taken in an adverse manner as corporate social responsibility is usually directly tied to a company's vision statement. Firms like fast food retailers or snack food manufacturers like McDonald's or Subway should have elaborated vision statements which might help them in creating sustainable growth curves, promoting environmental protection and furthering employee interests. For example, KFC employed differently abled servers in various countries. As a result, the group parent of the KFC, Yum! Restaurants were awarded the National Award for Empowerment of Persons with Disabilities. The sales of KFC in a very short span of time touched the sky and the company charted a path of its own. Corporate Social responsibilities go further beyond, where businesses begin to give back to the communities they have their footprint in and contribute their earnings with those around them (Figure 7.3).

Figure 7.3: KFC an award-winning food joint for CSR.

Source: http://1.bp.blogspot.com/-cEhI7bepCFQ/UX_lTGl36yI/AAAAAAAAAPI/JKWlE5QO6T0/s1600/CSR_KFC.jpg

7.5.3. An Initiative beyond Brand Equity

Many researches in recent days have given shocking figures which states that, around 8% of consumers feel that these organizations offering CSR are being proactive to protect the environment, as opposed to 44% among individuals, and 41% by non-profits. However, there have been another segment to the same kind of the research which shows that companies who demonstrate deep accountability in going above and beyond the mandated effort of corporate social responsibilities required by respective governments give them the absolute reason to connect with them. What is exactly a company for an individual? Answering this question: a company can be defined as an entity or an invisible proprietor or group of proprietors that engages in business transactions in order to achieve a particular goal.

In developing countries like India, there is a huge gap between urban-rural supply and also patronage gaps, where the people belonging to a rural area are usually at a disadvantage. An essential component of Indian food exports is biscuit manufacturing companies. These companies are not just launching innovative campaigns but are also thinking out of the box in order to build a more responsible organizational set-up which would have a futuristic outlook for the world. For example, a beach cleaning campaign conducted by Dunkin Donuts. Companies are also taking initiatives to create maximum number of employment opportunities for their neighbors, they are also trying to give fair prices for the acquisition of land and trying their level best to have synergies between the two.

For the sustenance of human life, food is very important. The big players in food market (which are valued in trillions of dollars) are doing good with the scale and also in numbers. Improving the image of the industry that not just uses valuable resources from a region to sell it at premium prices to first world countries without returning a share of the profits to the area is the responsibility of every player in that industry. The food manufacturing industry needs to invest a lot in certain sectors to have food exports that would bring huge reputation to the country (whether developing or developed), along with the increase in revenue. Companies which go beyond their usual corporate social responsibilities will reap more substantial rewards than currently imagined. For any food industry, the basic key to success in CSR for food supply chain is to have innovative initiatives which would help them build the brand equity along with the bucks. For example, Mother Dairy and KFC.

7.6. CONCLUSION

From the above discussion we can conclude that it is very important to build awareness and spread knowledge amongst all the food companies and specially amongst small and medium enterprises. These companies must be encouraged to have stated as well as documented policies, targets, and objectives. Having Corporate Social Responsibilities in the supply chain which is an application in the food industry, must be exercised in such a way that no harm is made to the environment and they should be encouraged to make specific commitments rather than general statements that cannot be implemented.

Making any false commitment would disappoint the potential consumers. Environmental practices, in the context of CSR, must be linked to costs and benefits in order to give complete justification to the investors who have invested for such initiatives. More systematic approaches to environmental CSR must be encouraged, such as Environmental Management Systems, because of the more technical and regulatory heavy nature of environmental issues. These companies also need more awareness for the wider environmental as well as social issues that can affect their operations given that social issues are not apparently visible and if visible then might not get tackled in an effective manner.

REFERENCES

1. Bakemate – Biscuit Exporter | Chocolate and Wafer Manufacturer, (2018). *The Importance of CSR in Food Manufacturing Industry – Bakemate – Biscuit Exporter | Chocolate and Wafer Manufacturer.* [online] Available at: http://bakemate.in/importance-csr-food-manufacturing-industry/ [Accessed 14 Mar. 2018].
2. Businesscasestudies.co.uk., (2018). *Corporate Social Responsibility*, Creating shared value in the supply chain – Nestlé | Nestlé case studies, videos, social media and information | Business Case Studies. [online] Available at: http://businesscasestudies.co.uk/nestle/creating-shared-value-in-the-supply-chain/corporate-social-responsibility.html [Accessed 14 Mar. 2018].
3. Favero, C., & Soriani, C., (2004). *Corporate Social Responsibility in the Food Industry*. Valuing the use of corporate social responsibility in the Australian food industry. [online] Dspace.unive.it. Available at: http://dspace.unive.it/bitstream/handle/10579/7442/840125–1173024.pdf?sequence=2 [Accessed 14 Mar. 2018].
4. Fritz, M., & Schiefer, G., (2008). *Sustainability in Food Networks*. [eBook] Available at: https://ageconsearch.umn.edu/bitstream/52648/2/fritz.pdf [Accessed 14 Mar. 2018].
5. Heikkurinen, P., Jalkanen, L., Matti, J., & Maaria, U. K., (2011). *Corporate Responsibility in the Food Chain: The Criteria and Indicators*. [online] Available at: http://centmapress.ilb.uni-bonn.de/ojs/index.php/proceedings/article/viewFile/1247/247 [Accessed 14 Mar. 2018].
6. Maloni, M., & Brown, M., (2006). Corporate social responsibility in the supply chain: An application in the food industry. *Journal of Business Ethics, 68*(1), pp. 35–52.
7. Maloni, M., & Brown, M., (2006). *Corporate Social Responsibility in the Supply Chain | Food Communication*. [online] Sites.jmu.edu. Available at: http://sites.jmu.edu/foodcomm/bibliography/maloni-and-brown-corporate-social-responsibility-in-the-supply-chain/ [Accessed 14 Mar. 2018].
8. Nagyová, Ľ., Holota, T., Košičiarová, I., & Holienčinová, M., (2016). *Corporate Social Responsibility in Food Manufacturing Companies – Environmental Dimensions*. [eBook] Available at: https://acta.mendelu.cz/media/pdf/actaun_2016064031037.pdf [Accessed 14 Mar. 2018].

9. Roberts, W. R., (1992). *Determinants of Corporate Social Responsibility Disclosure: An Application of Stakeholder Theory*. [eBook] Available at: http://repository.binus.ac.id/2009–1/content/f0122/f012243637.pdf [Accessed 14 Mar. 2018].

10. Yumcsr.com. KFC | Yum! Brands CSR Report. [online] Available at: http://www.yumcsr.com/brands-overview/kfc-page.asp [Accessed 14 Mar. 2018].

CHAPTER 8

FUTURE OF FOOD INDUSTRY

"Achieving the Global Goals for Sustainable Development will require a new level of insight on how uncertain futures might unfold. These scenarios are a powerful tool for leadership action for 8.5 billion people to access sustainably-produced nutritious food in 2030."

—David Nabarro, Special Adviser on the 2030 Agenda for Sustainable Development and Climate Change, United Nations

CONTENTS

In this chapter, we will discuss about the future of the food industry, what are the potential scenarios for the future world and appraising the new technologies in the food system. Next, we will discuss on the important drivers of change affecting the food system and emerging global trends in food sector. In the last part of the chapter, we will focus on the next generation food systems-like food consumption with eco-friendly practices, urban farming and Lab-grown foods, the internet of food and various other practices.

8.1. INTRODUCTION

Twenty years from now, the way food is produced, sold, and consumed in OECD (Organization for Economic Co-operation and Development) countries will probably suffer a drastically change. Projections of food supply and demand, suggest substantial transformations in consumption patterns, in technologies, in policies, and in international trade. As a result, market organization in OECD countries is likely to evolve rapidly. Some of the important drivers of change affecting the food system are discussed. Finally, some of the major trends in the future organization of the agro-food industry, including differentiation, concentration, and vertical co-ordination. "Food technology is the science that compact with all technique and activities involved in preserving, processing, and manufacturing the food stuff."

8.2. FOUR POTENTIAL SCENARIOS FOR THE FUTURE WORLD

The market for food industry is very uncertain, the matrix below in the figure describes four different scenarios for the future of global food systems. The analysis has been made on the basis of different assumptions and these four scenarios are based on those assumptions. The scenarios are further described on the basis of implications which would help in shaping a better world for tomorrow. The four scenarios analyzed to deal with few uncertainties are:

Survival of the Richest

The world full of resources and it has resource-intensive consumption with equally disconnected markets. The global economy is running at a slow pace where there is a stark division between the "haves" and "have-nots" for the consumption of resources.

Unchecked Consumption

The global market is interconnected and is a resource-intensive consumption place. GDP is high which is inclusive of high environmental cost. Therefore, there comes a point where the consumption of resources remains unchecked.

Open-source Sustainability

A market which links you with the future and its resource efficient has helped in increasing the international cooperation and innovation in the food sector.

Local Is the New Global

In many countries, the market for food sector is fragmented especially in local markets and have resource-efficient consumption. Countries which are rich in its resources focus more towards their local foods, whereas import-dependent regions become hunger hotspots (Figure 8.1).

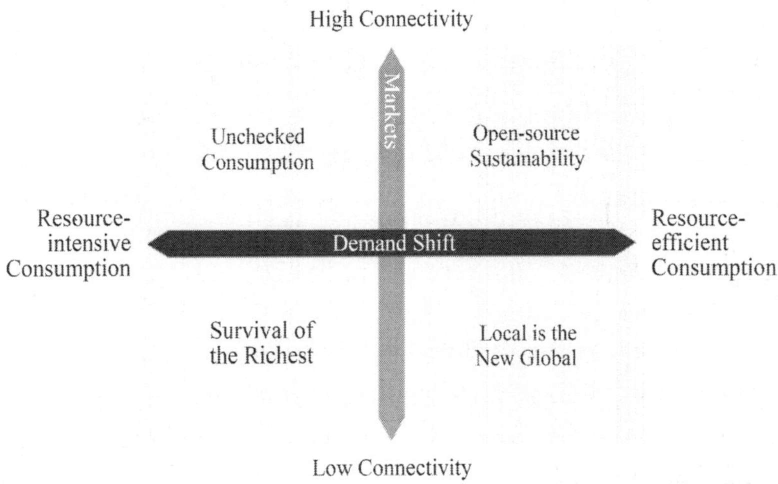

Figure 8.1: Scenarios and its implications for a better future world in the food industry.

Source: http://www3.weforum.org/docs/IP/2016/NVA/WEF_FSA_Futureof-GlobalFoodSystems.pdf

8.2.1. The Implications for the Scenarios to Build a Better Future

There are many implications required to shape the better world especially in case of the food sector. The implications are basically meant for various scenarios as discussed above. Those implications are discussed below in this section. Any of these scenarios can take place which could be an early sign out of all four scenarios present in our world today. Anyone scenario out of the four could become a reality by 2030. When combined together, these scenarios illustrate that today's food systems require a fundamental transformation to meet human needs within planetary territories in 2030.

Various insights are required for such situations. Some of those insights are:

1. **Consumption of resources have the power to make or break global health and sustainability**

   This scenario emphasizes more towards the importance of motivation which would enable and encourage consumers to eat more resource-efficient diets in their respective contexts.

2. **In order to have nutritious and sustainable food on every plate, it is important to redesign the food production systems**

   Such transformation in food production system would bring assurance in context to quality and not quantity of agricultural production.

3. **Changes in climate affects all the four future scenarios and poses a significant threat to the environment**

   Climate change and degradation of natural resource might compromise the long-term productive capacity of food systems, compromising social stability and economic well-being.

4. **Food system dynamics are expected to intensify inequality within and between nations**

   Growing inequality will affect the future of the food industry at a global level. Each scenario has its own winners and losers; the disparity between them is most evident in a disconnected world of more resource intensive demand.

5. **Fourth Industrial Revolution technologies and other innovations can revolutionize food systems but will introduce new challenges**

Technology innovations will dramatically reshape how we produce, manage, and demand food in select markets, but their effects will be unevenly distributed.

The analysis recognizes opportunities for leaders to pursue food systems transformation:

1. A new era of business in the food sector will help in capturing market opportunities so that investment could be made in health and nutrition and will contribute to greater resiliency in global markets. The improved market system would increase the resource efficiency of business operations, and leverage technology to address social and environmental challenges in food systems.
2. New, bold "smart policies" are highly required to redesign food systems, which will bring opportunities to adopt a "whole of government" approach to integrate the true costs of food systems; link food, agriculture, and environmental policies to healthy diets; and create an enabling environment for inclusive technologies.
3. Another important characteristic which should be considered while redesigning the food systems is keeping social and ecological priorities at the center of redesigned food systems. This would include efforts to address structural inequality and will help in meeting the basic needs of the consumers. It will also influence new dietary norms and aspirations and will elevate the needs of future generations. Responsive and responsible leadership is needed from all sectors to secure inclusive, sustainable, efficient, nutritious, and resilient food systems. Our choices through action or inaction will determine our path into an uncertain future.

8.3. IMPORTANT DRIVERS OF CHANGE AFFECTING THE FOOD SYSTEM

This is an exclusive period in history – choices and conclusions made now and over the next few decades will excessively impact the future:

- For the first time, with the number of people leveling in the range of eight to ten billion towards the middle of the century or in the two decades that follow, there is now a high likelihood that growth in the global population will cease.
- One of the most dominant factor of human activities are the earth system: "decisions made now to mitigate their detrimental effects

will have a very great influence on the environment experienced by future generations, as well as the diversity of plant and animal species with which they will share the planet."

- There is now an emerging worldwide agreement, embodied in the "Millennium Development Goals," which says that there is an obligation on everybody to try to end hunger and poverty, whether to the poor in more wealthy nations or in low-income countries.

Over the next 40 years, threats from interacting drivers of change will converge in the food system. It is very indispensable to have a careful valuation of the implications of these drivers if major pressures are to be predicted, and future risks are to be managed. Some of the most important drivers are outlined below:

Global population increases

The overall population is increasing all around the world, this increase can mostly be seen in low and middle- income countries, for example, Africa, etc. I assume the population today is around seven billion then most likely it will rise to around eight billion by 2030 and maybe it will be over nine billion by 2050. For example, the population of Africa is projected to double from one billion to approximately two billion by 2050. However, population projections are uncertain and will need to be kept under review.

Factors such as educational attainment, GDP growth, access to contraception and gender equality; possibly the single most crucial factor is the extent of female education are some of the major factors which are affecting the population.

With the rise in the number of people moving from rural areas to cities that will need to be serviced with food, water, and energy, their population growth will also combine with other transformational changes, particularly in low- and middle-income countries.

Changes in the size and nature of per capita demand

Some food items (such as grain-fed meat) require considerably more resources to produce than others, therefore dietary changes are very significant for the future food system. However, forecasting patterns of dietary change is multifaceted because of the way pervasive cultural, religious as well as social effects interact with economic drivers.

- **Meat:** By the middle of the century the meat consumption is expected to increase in per capita consumption (kg/capita/annum) to 52 kg which is 32 kg at present. Consumption is nearing a plateau in high-income countries. Whether consumption of meat in chief economies such as China and Brazil will stabilize at levels similar to nations such as the United Kingdom, or whether they will increase further to reach levels more similar to the USA is extremely indeterminate. However, there are chances of serious implications for competition for land, water, and other inputs with major increases in the consumption of meat, particularly grain-fed meat, this will also affect the sustainability of food production.
- **Fish:** At least in line with other protein foods, and particularly in parts of east and south Asia the demand for fish is expected to increase substantially. By further expansion of aquaculture, the majority of this extra demand is anticipated to meet, which will have substantial consequences for the organization of aquatic habitats as well as for the supply of feed resources.

Major uncertainties around future per capita consumption include:

- The degree to which consumption will increase in Africa.
- The degree to which diets will meet on those typical of high-income countries today.
- Whether local differences in diet (chiefly in India) continue into the future.
- The degree to which increased GDP is associated with reduced population growth and bigger per capita demand – the detailed nature of how these dissimilar trade-offs grow will have a major outcome on gross demand.
- Future governance of the food system at both national and international levels

There are numerous features of governance which have a substantial effect on the workings of the food system:

- Over recent decades the globalization of markets has been a chief subject shaping the food system and the degree to which this continues will have a considerable consequence on food security.
- The rise and constant development of new food superpowers, particularly India, China, and Brazil. Russia is already significant

in worldwide export markets, and there is possibility that it will even grow more in coming years, with a large supply of underutilized agricultural land.

- The leaning for alliance in the private segment, with the advent of a limited number of very large international corporations in the fisheries sector, in agribusiness as well as in the food processing, distribution, and retail sectors. With the entry into international markets of new companies from emerging economies, there are chances and evidence that the trend may get reversed.
- Making subsidies, trade limitations and other market interferences already have a chief consequence on the global food system. How the future will be developed is crucial.
- The degree to which governments act individually or jointly to face upcoming challenges, mainly in communal resources, trade, and volatility in agricultural markets. The inadequate governance of transnational fisheries, despite severe resource and market pressures, illustrates in microcosm numerous of the institutional and political difficulties to effective shared action.
- The adequacy of the existing transnational institutional planning to respond to future intimidations to the universal food system, and the political will to allow them to function successfully, is unclear.
- The control of increasing areas of land for food production (such as in Africa) will be influenced by both past and future land-purchase as well as rental contracts – including both sovereign wealth funds and business.

8.4. EMERGING GLOBAL TRENDS: FORCES OF CHANGE IN THE FOOD SYSTEM AND DISRUPTION

The main objective of food industry is to have a sustainable, healthy, inclusive, nutritious, and efficient food systems, which is not easy to achieve. There are various challenges and also opportunities for food systems in the future to achieve the above-stated objectives. Some of the emerging trends which would allow the food system to change have been discussed below with few disruptions as well:

- **Demographic Shifts:** Demographic shifts in the food sector are increasing and changing the demand for food amongst the consumers. The world population is expected to reach 8.5 billion by 2030 and the size of the global middle class is projected to increase from 1.8 billion in 2009 to 4.9 billion by 2030. The growing size of smart cities and rate of urbanization are further intensifying the pressure on food systems; it has been projected that, by 2050, over two-thirds of the global population would move towards the cities.
- **Macroeconomic Trends:** The increasing inequality of wealth and slow economic growth are two economic challenges encountered by food systems in the marketplace. Within the agricultural sector, there are around 800 million people who live below the global poverty line, which represents nearly four-fifths of the world's population approximately. The world is highly dependent on small agriculture producers: some 500 million smallholder farms worldwide produce over 80% of the food consumed in the developing world. At a larger scale, the richest 1% of the global population currently have more wealth than the rest of the world combined. This economic inequality is increasing day by day, translating into slower growth and compromising social cohesion.
- **The Triple Burden of Malnutrition**: Undernourishment, micro-nutrient deficiencies, and over-nutrition presents serious challenges and might get more worst if effective steps are not taken to tackle these situations. Around 800 million people sleep hungry and, although significant progress has been made, but still developing countries are fighting with these problems at a larger scale. Over 2 billion people suffer from micronutrient deficiencies, which can lead to diseases and develop challenges, such as stunting. Meanwhile, over 2 billion adults are overweight or are suffering with obesity. Poor nutrition and health habits can lead to non-communicable diseases (NCDs), which are now the leading cause of death in all regions except Africa. The economic burden and costs of NCDs radiate through households in the form of lower wages and savings. These strain health and welfare systems, impacting business and the economic growth of countries.
- **Natural Resources:** Due to unsustainable agricultural practices,

there has been a lot of depletion in the natural resources. The food sector accounts for 70% of water withdrawal and agriculture, forestry, and other land use accounts for almost a quarter of global greenhouse gas emissions (of which half comes from land conversion).

Water withdrawals have increased threefold over the past 50 years and demand is expected to rise by a further 40% by 2030. About a third of all arable land is degraded – a figure that continues to grow and that is related to farming intensity. According to a report published by International Panel on Climate Change, "Despite efforts to decelerate climate change, the plans of 170 nations to curb emissions would still lead to an estimated temperature rise between 2.7°C and 3.7°C by 2100 – far above the critical 2°C global target, with resulting increases in food prices by as much as 84% by 2050."

- **Geopolitical Dynamics:** Geopolitical dynamics in the food systems are bringing food insecurity and other challenges. The world is currently experiencing the highest levels of displacement on record, with 65.3 million forcibly displaced people worldwide. At the same time, emerging political movements, and elections in the United States and Europe, are representing nationalist and isolationist tendencies which might affect trade agreements and collaborations between two or more countries, which is definitely a huge breakdown for this sector.
- **Existing Innovations and Emerging Technologies**: The emerging technologies and existing innovations in the food systems provide opportunities to tackle the challenges in an effective manner. But strategies adopted to tackle the challenges will give rise to new questions and the impacts will not be evenly distributed. In complement to existing technologies, like mobile platforms, new Fourth Industrial Revolution technologies will profoundly change the way today's generation live, work, and relate to one another. Food systems will be dramatically influenced by this new era and the new tools it offers, including artificial intelligence, bio-innovation, gene editing, big data, machine learning and robotics.

These innovations will continue to create unexpected amounts of data. Indeed, 90% of the world's data has been created in the past two years and is expected to continue to grow exponentially. Such technologies are majorly dependent on the internet through which industries get a chance to get connected with their consumer. The number of internet users quadrupled from 2005 to 2015 to over 40% of the world's population over 4 billion still don't have access and the speed of connectivity is slowing down. Gender inequality complexes this challenge because women in rural areas do not have access to the internet than men in the same communities.

8.5. APPRAISING NEW TECHNOLOGIES IN THE FOOD SYSTEM

- Though there is a need to respect the views of people who take a contrary view, innovative technologies (such as the use of cloned livestock and nanotechnology and the genetic modification of living organisms) should not be excluded a priori on ethical or moral grounds.
- Investment in research on modern technologies and skills is indispensable in light of the magnitude of the challenges for food safety and security in the coming years.
- There is a rigorous need of human and environmental safety to be established before its deployment, with open and clear decision-making.
- In the context of competing risks, decisions about the acceptability of modern technologies need to be made (rather than by simplistic versions of the precautionary principle); the potential costs of not utilizing and applying novel technology must be taken into account.
- Modern technologies may modify the connection between commercial benefits and food producers, and while designing governance of the food system this should be taken into account.
- To address food security there are multiple approaches, and there is a lot that can be done today with existing knowledge. Research portfolios need to include all areas of science and technology that can make a valued impression.
- Appropriate modern technology has the possibility to be very valuable for the deprived people in low revenue nations. It is very significant to integrate likely beneficiaries in decision-making at all phases of the development process.

8.6. FOOD SYSTEMS AND THE NEXT GENERATION

The impact of technology is not less on the food industry. Below, four technologies have been discussed in detail that may have a profound effect on how consumers might eat in the future:

Fully or partially automated vertical farms

Vertical farming has a huge effect at what some individual eats. By assembling farms on top of another and using automation, vertical farms can produce crops 100 times more effectively per acre when compared with usual conventional agricultural techniques.

If automated vertical farming gets introduced, then it would grow crops at twice the speed as usual, while using 40% less power, having 80% less food waste, and using 99% less water than outdoor fields. But, after all these relaxations, the problem for vertical farms is the high cost and it is still not clear when they will be viable on a commercial basis.

Introduction of Aquaponics in the agricultural sector

Another technology which has got introduced and is bring positive changes in the food sector is Aquaponics. Aquaponics is a unique combination of fish farming (aquaculture) with hydroponics (a subset of hydroculture).

Fish convert their food into nutrients which can get absorbed by the plants and these plants clean the water for the fish. Compared to conventional farming, this technology uses about half of the water, and increases the yield of the crops grown. An aquaponic system depends on different live components to work successfully. The three main live components are plants, fish (or other aquatic creatures) and bacteria.

In Vitro Meats

Meat is an expensive product and is resource-intensive in nature when it comes to the production of meat. For example, to produce one pound of beef, it takes 1,847 gallons of water. In vitro meat is sometimes referred to as cultured, synthetic, or cell-cultured meat.

To solve this problem, there is a technique called In vitro meats. It involves injecting muscle tissue from an animal into a cell culture, which cells to "grow" outside the animal's body. Interestingly enough, market demand seems to be there: one recent study found that 70.6% of consumers are interested in trying lab-grown beef.

Artificial Animal Products

One other route to get artificial meat is to use machine learning to grasp the complex chemistry and textures behind these products, and to find ways to replicate them. This has already been done for mayonnaise – and it's in the works for eggs, milk, and cheese as well.

8.7. TASTING THE FUTURE OF FOOD

As these new technologies scale and hit markets, the future of food could change drastically. Many products will flop, but others will take a firm hold in our supply chains and become culturally acceptable and commercially viable. Certainly, food will be grown locally in massive skyscrapers, and there will be decent alternatives to be found for both meat or animal products in the market.

With the global population rising by more than a million people each week, finding, and testing new solutions around food will be essential to make the most out of limited resources.

Johan Rockstrom, professor of environmental science at Stockholm University, once stated that it is high time to think about the implications of society's next 10,000 years by focusing on what happens within the next 50. "Food is the deciding factor." He also stated that, "Food is the primary cause behind loss of diversity, the largest user of fresh water, and it is responsible for 30% of all greenhouse gas emissions. If we get it right on food, we are likely to get it right for planet Earth."

8.7.1. Food Consumption with Eco-Friendly (Hi-Tech) Practices

A revolution in food industry will be highly technical and would involve various technologies to deal with all kinds of shortcomings. Many startups are busy with monthly innovations to bring a change in this sector. Startups like, The Food Assembly is making efforts to connect local farmers and food producers with consumers. This will allow farmers sell their produce in an ethical marketplace, on the other hand, food bag tackles environmental waste with a composite packaging solution. Platforms such as Too Good To Go and Winnow are also making a difference by creating easy interfaces to help users manage their food waste and Modern Meadow uses bioprinting to grow meat and leather in labs. Scientists and entrepreneurs are putting the ideas out there and investors are taking note. Thankfully, the future of our food looks to be in safe hand.

8.7.2. Urban Farming and Lab-Grown Foods

On-demand farming services are helping cities become self-sufficient in their food production and are also trying to eliminate waste and reduce their environmental impact at the same time. In London, the aquaponics enterprise, GrowUp Urban Farms, produce fish, salads, and herbs in unused city spaces to sell them in a wholesale market. In Berlin, Infarm has implemented its hydroponics system in local supermarket stores. "The latest addition in this list is of Agrihood, which is a space which feeds 2000 people in America for free through a three-acre project which includes two acres of growing land as well as a fruit orchard with 200 trees."

In case of urban farming, the intensity of it means that crops grow quicker and requires less water in the process. The main goal at present is to take the teachings of GrowUp, Infarm, and Agrihood and put them to use in developing countries where food sources are entirely dependent on harsh climatic conditions.

8.7.3. The Internet of Food

The global online grocery market is set to grow to 80 billion pounds by 2018. At one side of the market, online ordering and delivery infrastructure model is growing at a rapid pace, without realizing that the actual journey food has to go through from farm and then to consumer which is still slow and is full with obstacles. To address this issue, many people from this sector including technical sector are joining their hands to work for an Uber-style redesign of existing systems, in order to provide a new way to connect customers and growers. Jennifer Goggin, co-founder, and CEO of Farmers Web explained the entire procedure in a precise manner by stating that: "Like Uber's on-demand transportation model, this missing link could come from an online logistics platform that would allow farms to utilize existing transportation for their products—e.g., via a neighboring farm that is already making their own deliveries but has extra room their truck. A powerful and widely-used technology platform is key to making this vision a reality."

8.7.4. Cannabis Digestible

2017 has been declared as the year of the cannabis industry. The Pantone's 2017 announced "Greenery" as its Color of the Year. Cannabis edibles has taken up a huge portion of the industry's growth. With cannabis cookbooks, vodka, wine, crisps, water, and more already in the market, brands are now reaching for innovative tech solutions for cannabis extraction and are trying their hands on into the manufacturing of cannabis digestible to lure customers with innovations. A fresh example for cannabis edibles are: chocolates (Defonce Chocolatier), online marketplaces (Mirth Provisions), coffee (K Cup coffee), and subscription boxes (Nug Run), as well as the AmeriCanna cannabis cafe which just landed $1 million in investment (Figure 8.2).

Figure 8.2: Defonce Chocolatier.

Source: https://www.youthvariety.in/wp-content/uploads/2018/02/defonce.jpg

8.7.5. Delivery

In 2016, an extended research was conducted by Morgan Stanley which concluded that the food delivery industry is currently valued at $30B, but has a whopping $210B worth of potential to uncover in the years to come. It's not so surprising that everyone wanted a share of the delivery action in 2016, including Facebook. Facebook when realized this revolutionary change in the food sector, launched its Delivery button in the summer and Google came up with the Google Maps update which allowed users to place local delivery orders using services such as Seamless and Grubhub.

People now expect a "Drone Delivery." Delivery by drone is still very much in its testing phase but not too far. Google has been testing Chipotle burrito delivery by drone at the Virginia Tech campus. In Singapore, Food panda also tested drones in an aim to reduce delivery times. Eventually, it intends to expand the concept across markets including Singapore, Malaysia, Hong Kong, India, and Thailand.

Meanwhile, in London, food delivery robots are already roaming the streets, delivering restaurant meals for ten Just Eat pilot restaurants. The bots have cameras fitted and require a code to access their contents.

8.7.6. Performance Dining

Performance dining or 'neurogastronomy' is one of the fastest growing trends curated on the belief that experience, entertainment, and information should be combined to build new relationships between people, materials, and eating. In the past, this manifested itself in elaborately-staged half-dinner, half-theatre setups (Figure 8.3).

Figure 8.3: Performance dining at grand canyon hotel.

Source: http://www.grandcanyongrandhotel.com/assets/dining-john-smith.jpg

But in the future, gastronomy, and performance dining will be used more frequently with the help of technology to provide customers with data-based services and emotional experiences beyond the initial taste of food and drinks. In Asia, robotic chefs and waiters save on staff wages and a San Francisco-based chain of restaurants called Momentum uses a robo-cook.

Apart from the cooking part, another real area to watch out for, is 3D-dining. Food Ink are one brand running some of the first 3D dining pop-ups around Europe at the moment. Self-described as a "conceptual pop-up dinner series where fine cuisine meets art, philosophy, and tomorrow's technologies," guests are invited to sit at 3D-printed chairs and tables before a seven-course menu is printed with a robotic arm, right in front of their eyes. Food Ink hopes to "serve as a platform for a public conversation about how these emerging technologies are rapidly challenging and changing the way we eat, create, share, and live." (Figure 8.4)

Figure 8.4: Food ink: A 3D printing restaurant.

Source: https://i.ytimg.com/vi/xk9hNin_sag/maxresdefault.jpg

REFERENCES

1. Bbc.co.uk, (2014). *Food Packaging and Labeling*. [online] Available at: http://www.bbc.co.uk/schools/gcsebitesize/design/foodtech/packaginglabellingrev5.shtml [Accessed 17 Mar. 2018].
2. De Wilde, S., (2016). *The Future of Technology in Agriculture*. [eBook] Available at: https://stt.nl/stt/wp-content/uploads/2016/05/ENG-Toekomstverkenning-agri-food-Web.pdf [Accessed 17 Mar. 2018].
3. Lee, C., (2017). *What is In Vitro Meat?* [online] Office for science and society. Available at: https://www.mcgill.ca/oss/article/technology-health-and-nutrition/what-vitro-meat [Accessed 16 Mar. 2018].
4. Lewin, K., (2017). *Food Trends 2017: How Technology is Shaping Our Future*. [online] Creativepool.com. Available at: https://creativepool.com/magazine/technology/food-trends–2017-how-technology-is-shaping-our-future.11960 [Accessed 16 Mar. 2018].
5. *Shaping the Future of Global Food Systems: A Scenarios Analysis*, (2017). [eBook] Available at: http://www3.weforum.org/docs/IP/2016/NVA/WEF_FSA_FutureofGlobalFoodSystems.pdf [Accessed 16 Mar. 2018].
6. Shapiro, K., (2016). Pantone picks the vibrant shade of cannabis as its color of the year for 2017. [online] *The Cannabist*. Available at: https://www.thecannabist.co/2016/12/09/2017-pantone-color-of-the-year-greenery-cannabis-style/69108/ [Accessed 16 Mar. 2018].
7. *The Future of Food and Agriculture: Trends and Challenges*, (2017). [eBook] Available at: http://www.fao.org/3/a-i6583e.pdf [Accessed 17 Mar. 2018].
8. *The Future of Food and Farming: Challenges and Choices for Global Sustainability*, (2011). [eBook] Available at: http://www.saiplatform.org/uploads/Library/11–547-future-of-food-and-farming-summary.pdf [Accessed 17 Mar. 2018].
9. *The Future of Food Long-Term Prospects for the Agro-food Sector*, (1998). [eBook] Available at: https://www.oecd.org/futures/35391719.pdf [Accessed 17 Mar. 2018].
10. *The Future of Food: New Realities for the Industry*, (2017). [eBook] Available at: https://www.accenture.com/us-en/_acnmedia/PDF–70/Accenture-Future-Of-Food-New-Realities-For-The-Industry.pdf [Accessed 17 Mar. 2018].

11. Yada, Y. R., (2015). *Emerging Technologies and Innovation in Food Science and Technology: Meeting Consumer Needs of the Future.* [eBook] Available at: http://bio.albertainnovates.ca/media/77229/3-ifsfood2015-may26-rickey_yada.pdf [Accessed 17 Mar. 2018].

CHAPTER 9

CASE STUDY

"I probably spend more on food than a lot of people, and I feel good about the whole food chain I'm supporting when I'm doing it. But even I have to remind myself. I'm always complaining about the prices at the farmer's market."

—Michael Pollan

CONTENTS

Innovation is a process that goes beyond theoretical conception, through technical invention to commercial exploitation. Therefore, R&D is an element of innovation. One key distinction between R&D and innovation is that in the former, we don't necessarily know the outcomes. To create an innovation, the desired outcome must be known before tackling the problem with R&D. All research has a purpose. However, a commercializing purpose is very different from a knowledge-seeking purpose. Funding mechanisms to deliver in both areas will have to follow very different rationale. From my point of view, that grants investing in knowledge-seeking endeavors should be discussed separately from the discussion of a national innovation system. This case study highlights the sustainable innovation of the food supply chain and has been discussed in detail.

9.1. SUSTAINABLE CO-INNOVATION: THE FOOD SUPPLY CHAIN AS A CASE STUDY

9.1.1. Introduction

During 2003/04, Australians spent more of their salaries in food and beverages than in any other household expenditure category, including housing. Although the manufacture of food, tobacco, and beverages (FTB) contributes with almost 20% of the total value of manufacturing industries, the expenditure in research and development (R&D) of the FTB sector represents less than 8% of the total R&D investment in the manufacturing sector. It is fair to say that R&D is not a synonym of innovation. Therefore, R&D is an element of innovation. This view is supported by the recent report "Public support for science and innovation," which defines innovation as:

"Deliberative processes by firms, governments, and others that add value to the economy or society by generating or recognizing potentially beneficial knowledge and using such knowledge to improve products, services, processes or organizational forms. From the perspective of this study, these improvements may be specific to the entity, to the industry, country or world, and could be incremental or novel."

Historically, public R&D organizations have been an integral part of the Australian innovation system. However, it has become clear that these organizations alone are not capable of providing the full spectrum of activities required to commercialize inventions. One key distinction between R&D and innovation is that in the former, we don't necessarily know the outcomes. Therefore R&D endeavors are always a success, as they always increase knowledge. However, new knowledge may not deliver commercial outcomes, at least in the short term. To create an innovation, the desired outcome must be known before tackling the problem with R&D. Furthermore, innovation is fallible. As stated by DTI and DEFRA, "innovation can always be accurately targeted to a commercial outcome but unlike research it can fail."

9.1.2. What Motivates Scientists?

R&D organizations contribute substantially to innovation, by:

a) Increasing the amount of useful knowledge;
b) Training skilled graduates;
c) Creating new scientific methodologies;
d) Creating knowledge networks nationally and internationally;
e) Increasing scientific and technological capacity. And occasionally;
f) Creating new companies, often in collaboration with business-savvy partners.

Given that most Government innovation initiatives in the past have placed significant demands on universities and R&D organizations to commercialize their scientific production, it is worth exploring the reasons why young scientists choose to work in publicly-funded research. Ultimately, people, and not policies or institutions, are the innovators. Jindal-Snape and Snape performed a study to understand the motivators in 18 scientists in a government research institute in the UK. The findings in this study were that curiosity, the need to do good science and the desire to make a difference, are the predominant motivators for scientists. None of these motivators related to financial issues (e.g., bringing external income) or management aspirations (e.g., becoming senior managers in the organization). Bigliardi et al. found that other motivators in engineering-oriented personnel include having the opportunities and facilities to have pleasant social interactions, such as interest-group meetings during breaks or off-duty hours. The social dimension covers all contacts the employee has with other people, both inside the organization and in the environment of the organization – for example, customers, suppliers, colleagues, and managers. Factors related to this dimension are communication, feedback, feelings of solidarity, acceptance, leisure time, status, leadership, power, and the need to help others.

These observations do not aim to stereotype researchers as persons with no commercial drive and always resistant to commercialize the results of R&D. Indeed, a US study9 found that academic researchers who have research grants and contracts work more extensively with industry than those without grants or contracts. Furthermore, the scientists who have industry contracts interact with industry to a greater degree than those who are exclusively funded by governments. However, traditional academics and scientists do have concerns on the growing pressure on public R&D institutions and universities to produce immediate useful knowledge with close-to-market commercial benefits. This is often at the expense of longer-term programs and the development of basic scientific research, which are key ingredients for groundbreaking science. Recent studies show that researchers that maintain a funding relationship with a private sponsor (or a Government grant that is set to respond to specific demands of the private sector) experience a decrease in their academic publications by 25%. Furthermore, academic talent is not necessarily a funding criterion for sponsors10. Therefore, concerns over a decrease in traditional scholarly activities as Government funding focuses on reaching commercial goals are funded. This author knows of more than one public R&D researcher that has either left the organization to join a university to continue their scientific endeavors "with more freedom," or that has abandoned research to pursue a business career. Both types of researchers often leave in frustration, due to the complexities of following the double mandate of public R&D institutions: the public benefit driver vs the commercial driver. From the previous discussion, we conclude that a nation that excels at innovating does not necessarily excel at creating new knowledge and groundbreaking science. The opposite statement is also true: a nation that excels at developing new scientific frontiers does not necessarily excel at innovating. Do we want to see Australia as a nation of innovators or as a nation of Nobel laureates in the coming decades? Or do we want both? What is the purpose of the Australian Government?

9.1.3. Science in the Innovation Continuum

All research has a purpose. However, a commercializing purpose is very different from a knowledge-seeking purpose. If the new Government is seeking a balance between these, it should be aware that funding mechanisms to deliver in both areas will have to follow very different rationale. My view is that grants investing in knowledge-seeking endeavors should be discussed separately from the discussion of a national innovation system.

Past Government systems seemed to focus on the transformation of public R&D organizations into innovation "factories." This may well be what it is required, but the roots of public R&D organizations are closer to knowledge-seeking endeavors than to innovation. Public R&D organizations and universities will require a clear mandate on the balance between funding for curiosity-driven science and science targeted to bring commercial outcomes. A possibly more important issue is that the Government investigates what is the right funding split between the two. Figure 9.1 shows a conceptualization of:

(a) What the innovation continuum encompasses;
(b) How science can feed into innovation;
(c) How the market can influence science; and
(d) How scientific endeavors are successful even if they don't produce an immediate innovation.

In the proposed concept, we assume that all ideas that progress into an innovation continuum, either supported by private of public funding, are worth exploring. New theories or knowledge (either born out of curiosity/creativity or market-driven needs) lead to basic research, which in turn could lead to developing a close-to-market application, or not. A reason for the latter may be that present technology limits commercial application (e.g., Einstein's theory of relativity was developed in 1905, but its application for the development of global positioning systems occurred 68 years later). Or perhaps the phase of applied research shows that the new product/service/technology will not be able to compete in price with current substitutes. Or perhaps the innovation infringes other's patents.

There are several reasons why innovations fail. But in this context, it is the implementation, rather than the new idea, what fails. The model presented implies that, for every research that fails to deliver an innovation in a particular application field (e.g., electronics), there may be n different fields that may benefit from the same research (e.g., instrumentation). This multiplicative effect is further illustrated in Figure 9.1a. This effect explains why one of the ways to assess the impact of patented inventions is to quantify the citations of the original patent, either in the same area of application or in different areas.

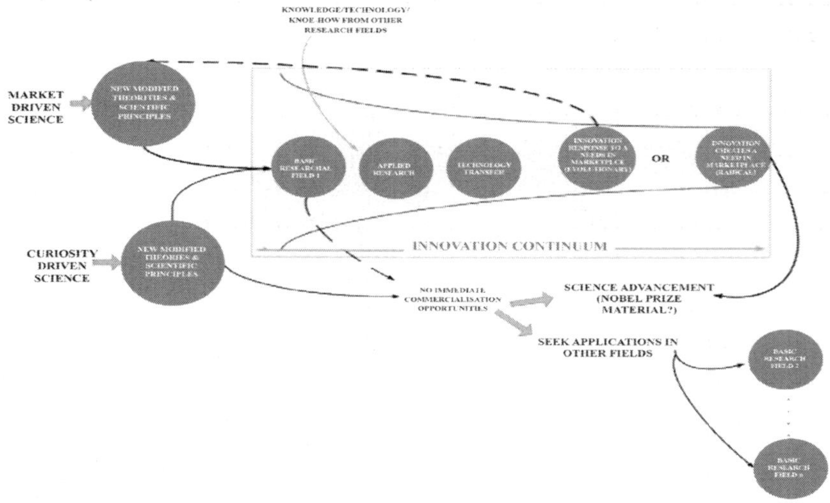

Figure 9.1: Model for innovation continuum and sources for innovation.

Source: https://dfat.gov.au/trade/topics/review-of-export-policies-programs/Documents/FoodChainInteeligencePartA.pdf

It is clear that some industries will benefit from a market push innovation model than from a technology push innovation model. In the following pages, I will focus on the food supply chains as an example of an area that would benefit from a market push approach. The application of concepts such as "innovation networks," "consumer-led innovation" and "forward commitment procurement" in an innovation framework for the food industry is also discussed. Furthermore, I propose the introduction of an innovation model based on these concepts and the use of consortia supplying innovations, buying innovations and creating the market conditions for the successful uptake of food innovations.

9.2. THE AUSTRALIAN FOOD SUPPLY CHAIN

The aim of the food industry is to transform agricultural raw materials into safe, convenient, good tasting and nutritious products for consumers, in a profitable and sustainable manner. To reach the consumers, all the participants in a food chain need some sort of integration that allows the flow of product until reaching the final user. Therefore, the food supply chain is defined as a set of interdependent companies that work closely together to manage the flow of goods and services along the value-added chain of agricultural and food products, in order to realize superior customer value at the lowest possible costs.

9.2.1. Structure and Drivers

Food value addition is generated by activities linked to primary and secondary processing, packaging, distribution, and retail. The Australian food industry is strongly driven by the following trends:

- **Horizontal issues:** Global trends in aging and health awareness, consumer trust and consumer satisfaction (convenience), food safety and traceability, financial sustainability related to costs of production/packaging/transport, innovation as an engine for growth and sustainability in production systems.
- **Vertical (specific) issues:** each link in the chain has particular concerns and drivers. However, some of these have a cumulative effect (e.g., acceptance of GMOs affects both growers and manufacturers).
- **Primary production:** In Australia, this sector encompasses a large number of relatively small firms, particularly upstream in the chain. In 2004, about 205,000 agricultural businesses (excluding services, forestry, and commercial fishing) were accounted for by the Australian Bureau of Statistics (ABS). About 130,000 of these businesses were farm households or farming enterprises. A large proportion of the remaining businesses include harvesting, primary processors and small road transport units.
- **Food manufacturing:** This is a diverse sub-chain, ranging from SMEs to major multinational companies. In 2004, about 9,000 food beverage and tobacco manufacturing businesses were accounted for by the ABS. The food manufacturing industry worldwide tends to be dominated by large, multinational firms and Australia is no exception, with the largest 50 food and beverage global corporations producing close to 75% of the domestic industry's revenue. The implication for innovation is that most R&D activities take place outside Australia, in the global R&D headquarters of the companies.
- **Food retailing:** This sector includes supermarkets, grocery stores, non-petrol sales of convenience stores at petrol stations, take-away food, and others (e.g., fresh meat, poultry, and fish; fruits and vegetables; liquor; bread and cakes, and specialized

food retailers). About 65,000 food retail businesses were accounted for during the 2004 survey of the Australian Bureau of Statistics. Food represents about 40% of the total retail sales and 70% of food sales are registered in supermarkets and grocery store retailing.

- **Food exports:** Australian farm exports were worth around $23.8 billion in 2005–06, accounting for 15% of total merchandise exports. Australia's profile as an exporter of rural-based products means that this sector's performance is affected by world commodity prices, exchange rates, trade barriers (e.g., tariffs and quotas), domestic support programs (e.g., production and export subsidies), and technical and regulatory barriers (e.g., product labeling and certification standards). Major export markets include Asia (Japan, Malaysia, Korea, Hong Kong, Taiwan), USA, and New Zealand. Australia is a net exporter of food, with an export surplus of $17 billion over food imports.

In 2005–06, the value of primary production goods almost trebled by the time the food reached the shelves in supermarkets. The largest increment in value occurred between primary and secondary production.

9.2.2. Global Competitiveness of the Australian Food Industry

A recent study by the European Commission compared the competitiveness of the European Union, Canada, Australia, New Zealand, US, and Brazil. The economic indicators included real value added, export share growth and real labor productivity. Australia's performance was found to be the weakest of all countries analyzed, as described in Figure 9.2.

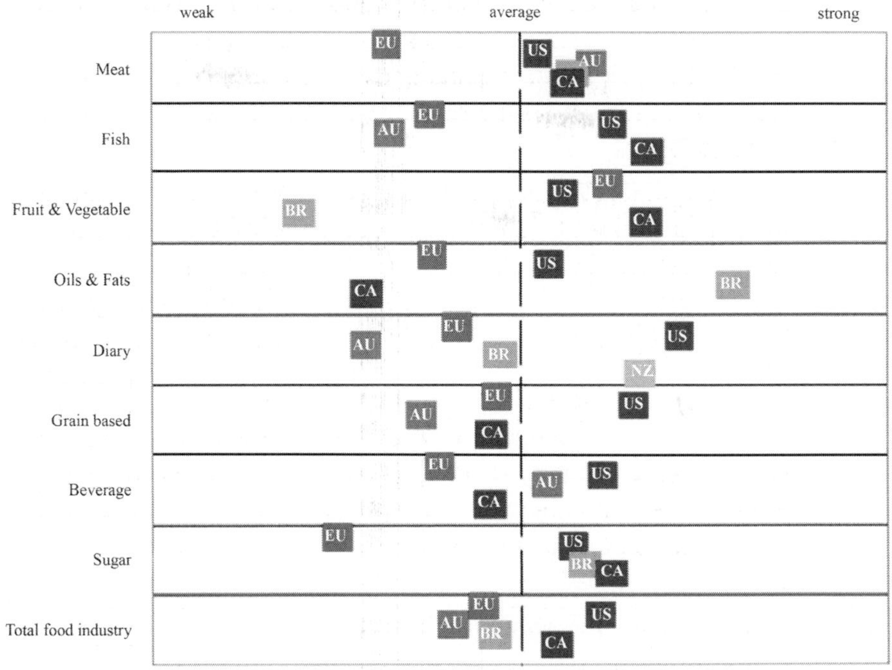

Figure 9.2: Global competitiveness of the Australian food industry.

Source: https://dfat.gov.au/trade/topics/review-of-export-policies-programs/Documents/FoodChainInteeligencePartA.pdf

9.3. INNOVATION STRUCTURES FOR FOOD SUPPLY CHAINS LED BY THE GOVERNMENT

It is difficult to quantify the impact of the historic public expenditure in innovation-related strategies on the modern Australian food industry. One reason is the concept of innovation itself, which encompasses a range of activities (e.g., new or improved products, business practices, and processes). These activities are difficult to track and assess through the use of the same performance indicators. Other reasons are the complexity of the industry and its eventful history, which evolved from fierce competition in the 1980's to relentless consolidation and the supremacy of retailers as a market power in present times. Therefore, the effect of innovation structures and funding strategies in the past years is likely to be confounded with the effects of other events, such as the economic outlook, consumption trends and many others.

Nevertheless, food industry experts have reached consensus in two issues:

(a) The Australian agri-food industry is in danger of losing markets that were taken for granted in the past (e.g., meat and grains), due to increased competition from countries such as New Zealand, South Africa, and China; and

(b) Innovation has a key role in stopping the decline in the performance of the food industry sector in Australia.

The expert consensus is better illustrated by the following statements:

"Ultimately, for Australian companies to remain competitive in the international marketplace, they will need to boost their commitment to R&D, continue to develop alliances with global brands and supermarkets, and be innovative in their product development and process technologies." (David McKinna, 2006).

"Posing challenges for some food manufacturers is the impact on category value and the viability of sustaining investments in branded products, including the level of innovation and R&D investment." (DAFF, Australia Food Statistics, 2006).

"The challenge for Australian farmers will be to identify and embrace new market opportunities and to drive competitiveness through innovation in production and processing, enhanced efficiency, improved economies of scale, attracting greater returns through better understanding of markets and changing consumer needs, through achieving greater equity through the supply chain, and through differentiating and developing specialized products." (National Farmers Federation, 2006).

"Product and process innovation will play a large part in the success that can be achieved. This will add convenience to a product solution rather than the traditional approach of supplying commodity lines in the form also available in retail stores." (Freshlogic, 2007).

"Food companies must continue to innovate and operate at high efficiency in order to achieve income growth." (Steve Spencer, Whitehall Associates, 2006).

"Innovation is the key to profitability in competitive world food markets." (NFIS, 2007).

9.3.1. Historic Investment in Food Innovation

This is illustrated by the 2004/05 statistics on national R&D expenditure. In 04/05, R&D investment in the primary production of foods was about $1 billion dollars, while R&D investment in food manufacturing was just below $300 million. This focus reflects just the opposite of the value chain, where the last stages of the chain (e.g., manufacturing and retail) are the ones that triplicate the value of primary production. Furthermore, the low levels of R&D in the food manufacturing stage are at odds with the knowledge-intensive character of the industry.

Most retail R&D is classified as manufacturing or services. There are no specific estimations of how much retail invests in R&D. Therefore, we assume that the food supply chain as a whole had an R&D public and private expenditure of about $1.3 billion during 04/05. Food manufacturing R&D expenditure increased from $192m in 2000 to $269m in 2005, as described in Figure 9.3.

The sectors with increased funding are (in decreasing order of gains in the past 5 years): others – mainly sugar, confectionery, prepared animal and bird feed and seafood (+$29 m), beverage, and malts (+$28m) and horticultural products (+$16 m). R&D funding for the dairy industry has decreased in about $2.5m per annum in the past 10 years.

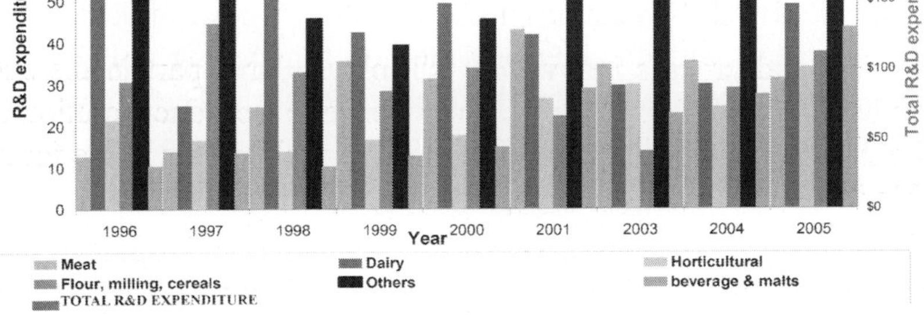

Figure 9.3: R&D funding for particular sectors in the food industry.

Source: Australian Food Statistics 2006; OECD 2004, Research, and development expenditure in industry.

Interestingly, these statistics do not necessarily encompass R&D in logistics and supply chain management. These activities are central to the operation of the food chain, yet firms operating in these interstices are not counted as part of the food industry and their innovation and R&D activities are accounted for in different industries. For example, R&D expenditure in retail logistics, transport, and storage is accounted as "services"; packaging R&D may be accounted for in manufacturing of chemicals, rubber, plastics, fabricated metal products and others.

9.3.2. Initiatives Supporting Food-Related Innovation

Two major Government funding initiatives for R&D closely linked to market demands were:

- The Supermarket to Asia (STA) Strategy from 1996 to 2002, which had as a main objective the expansion of Australia's agrifood exports to Asia. The Technical Market Access program was created through STA and it is still used by DAFF as a mechanism to enhance market access of Australian foods exports.
- The National Food Industry Strategy (NFIS) Ltd (2002–2007), an industry-led company, funded by the Australian Government to implement most aspects of the National Food Industry Council. The Food Industry Grants (FIG) were the most representative NFIS instruments for R&D funding.

The FIG Program had a budget of $34.7 million over five years. The Program provided support to Australian based food businesses for projects involving technical and scientific R&D. Projects funded needed to present strong prospects for commercialization.

Grants under the FIG Program were awarded on a competitive, merit-based assessment, with funding provided on a matching dollar-for-dollar funding basis for up to half of the eligible project costs. The majority of FIG proposals involved collaboration between Australian based food businesses and research organizations, including Food Science Australia and other CSIRO divisions, universities, and a range of technical consultants.

Data published by NFIS in regards to funded projects and grant applications from 2002 to 2004. A majority of these fall under 'processing,' followed by 'consumer value' (22% and 14%, respectively). The fruit and vegetable/horticulture products segment of the industry had the largest representation amongst FIG Program recipients (28%), followed by ingredients (22%).

The FIG program was highly successful and well regarded by all the participants. One particular weakness was the relative low participation of SMEs with respect to the participation of larger firms using the FIG grants to undertake research. Given the large percentage of foreign-owned/multinational firms, it is likely that FIG grants supported R&D activities that would otherwise be carried out outside Australia.

To analyze the split of public R&D funding, this is commonly categorized on the basis of the stage of innovation supported (e.g., basic research, applied for research, technology transfer).

- In general, CRCs tend to support heavily the primary production sector.
- The intention of CRCs seems to overlap with some RDCs; for example, there is one CRC for dairy, one RDC (Dairy Australia) and schemes such as the Geoffrey Gardiner Foundation in Victoria that also support dairy. For grains, there are two CRCs dedicated to this category, plus one RDC. There are other open funding sources (e.g., RIRDC) that are used to support grains and dairy projects to some extent. One may wonder if this duplication is a result of: (a) poorly managed R&D funding resources; (b) the needs of more R&D investment in particular industries; or (c) the diversity of the dairy and grains industries. Whatever the reasons, the duplication of areas for investment could lead to poor efficiency at the cost of lost opportunities in other industries.
- CSIRO supports various projects across the food supply chain through cofounding. Through Food Science Australia (a joint venture of CSIRO and the Victorian Government), funds for R&D in food manufacturing industry are available in the form of collaborative research contracts. However, available cofounding is directed to strategic areas for CSIRO/FSA (e.g., little emphasis is placed on retail innovations).
- DPIs predominantly support the primary sector, mainly in pre-production and production issues. State-led initiatives have a strong role in the industries relevant to each territory.
- There is a scarcity of funding available to investigate the final stages of the food value chain. There is also a lack of funding sources devoted to create innovations in non-R&D related fields (e.g., business practices, supply chain management, competitive intelligence).

- NFIS had a strategic role in supporting business innovation, manufacturing, supply chain and value addition projects. Additionally, NFIS approached food innovation with a 'whole of-the-chain' perspective. The termination of this initiative leaves a vacuum in R&D focused to food manufacturing and other areas of knowledge required to introduce food innovations in the marketplace (e.g., supply chain, consumer science, and packaging).
- Other food innovation programs are fragmented and do not allow a 'whole-of-the-chain' perspective (e.g., the Regional Food Producers Innovation and Productivity Program). However, trends in the global food marketplace indicate that entire supply chains formed by alliances between two or more players (as opposed to lone companies) will be required to compete in the food market. This is also true in other industries.

The high level of competition between public and private R&D providers for public funding. This brings the following disadvantages:

1. Wasted time and effort of R&D providers (public and private) in applying for a finite amount of public funding from RDCs and CRCs. R&D providers cannot always recoup the money spent in failed applications.
2. The competition process is not enhancing collaboration between universities and public R&D organizations. Neither is a process that improves communication between R&D providers and food supply chain players. Given the limited pool of funding, competing agencies prefer to apply as sole providers rather than sharing the funding with potential collaborators.
3. In many competitive calls for proposals from RDCs and CRCs, there is a lack of transparency as to how the proposals are evaluated. In addition to this, the timeframes between submitting an expression of interest and receiving the outcome of a full submission varies between 3 to 6 months. However, representatives of the food industry expect innovation projects to produce results in less than 1.5 years.
4. The competitive process does not support open innovation.

9.4. NFIS FOOD INNOVATION MODELS DURING 2002–2007

Figure 9.4 presents the food innovation network model proposed by NFIS. The model is self-explanatory, indicating the role of NFIS as a hub coordinator of the activities between the network partners.

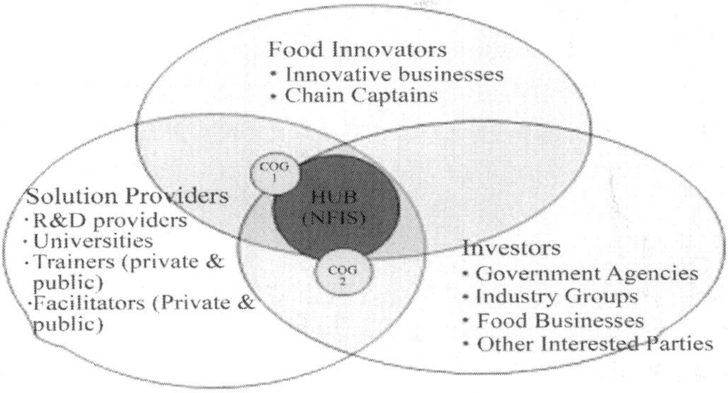

Figure 9.4: NFIS food innovation network model.

Source: https://dfat.gov.au/trade/topics/review-of-export-policies-programs/Documents/FoodChainInteeligencePartA.pdf

9.4.1. Centers of Excellence (COEs)

The two existing COEs were created by the National Food Industry Strategy in 2003, to respond to the decreasing R&D funding in the food industry. in the areas that were the most important innovation trends in the early 2000's: functional foods and food safety. The Centers of Excellence operated as virtual hubs of innovation, as illustrated in Figure 9.5.

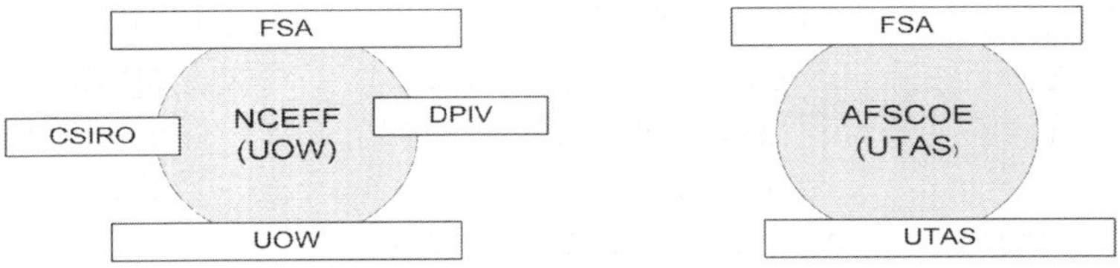

Figure 9.5: Centres of excellence created by NFIS in 2003. Notes: FSA= Food Science Australia, a joint venture of CSIRO and DPIV; UOW= University of Wollongong; UTAS= University of Tasmania.

Source: https://dfat.gov.au/trade/topics/review-of-export-policies-programs/Documents/FoodChainInteeligencePartA.pdf

Figure 9.6 shows the COEs model, where a virtual network of consortia partners agglomerate around one topic (i.e., functional foods, food safety) and provide a wealth of services to the industry. The review found that the COEs lacked formal mechanisms of engagement with industry: Why industry players should be involved? What is the cost of involvement? What is the motivation? The lack of engagement, the lack of a strong value proposition, weaknesses in governance and the duality of roles of the managers and scientists involved (e.g., CSIRO-centre view, as opposed to NFIS-centered view) were highlighted in a critical review undertaken by NFIS in 2005. Recently, AFSCOE (the most promising COE) has changed its affiliation to the Food Safety Centre, a self-sustained food safety information portal. The extent of funding from the host organizations (i.e., FSA and UTAS) to support the Food Safety Centre is unknown. In regard to the NCEFF, it is not known if it is still active.

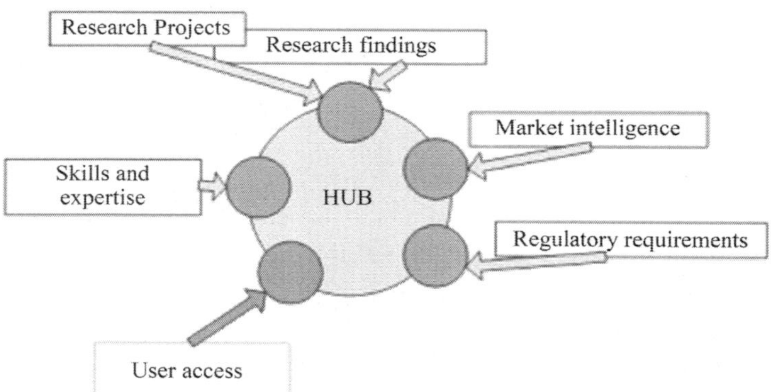

Figure 9.6: Hub model used for the NFIS funded centers of excellence.

Source: https://dfat.gov.au/trade/topics/review-of-export-policies-programs/Documents/FoodChainInteeligencePartA.pdf

9.5. CHALLENGES FOR PUBLIC ORGANIZATIONS FOCUSING ON FOOD RESEARCH

The analyses revealed the following limitations for the development of food innovations by public R&D organizations:

1. Although the levels of R&D expenditure in the Australian food industry are increasing, R&D funding is decreasing in areas that are heavily staffed in public R&D organizations (e.g., meat, dairy, cereals).
2. Supplier issues are basically a human resource issue. In the current job market, finding, and sourcing staff with the right mix of business and scientific skills required for public R&D is difficult. Particularly if those skills are in high demand in industries offering better employment conditions. An example of this situation is hiring scientists with expertise in industrial ecology or energy.
3. The role and influence of supermarkets (buyers) on the R&D agenda of manufacturers should be considered: The profitability of the retail sector is highly reliant on in-depth customer knowledge and reliable and efficient supply chains that can deliver the food products. Therefore, retail-led innovations focus on improving supply chain aspects. Additionally, the current environment for manufacturers (i.e., increasing productions costs and decreasing sell prices) means that there are fewer margins available for innovation and capability development.
4. The commercialization requirements for R&D providers can vary significantly. For example, RDCs have as a key priority the development of research that benefits the wider industry. This position does not encourage the use of intellectual property (IP) as a competitive advantage for a particular company. In the other hand, the industry normally expects full ownership (or full entitlement of licenses) of the IP developed with an R&D provider. Therefore, R&D providers need to have several strategies and business/IP models that accommodate these situations.
5. The levels of rivalry in the industry are high, as public food R&D organizations compete for decreasing industry and government funding with universities and private providers. Universities, in particular, are now competing in applied research grants, in addition to their traditional role of basic research providers.

9.6. POTENTIAL INNOVATION MODELS FOR THE FOOD INDUSTRY

I believe that open innovation models that:

(a) bring together the necessary roles and players for innovation;

(b) 'connect the dots' between the key innovation players; and

(c) create the necessary market conditions for the commercialization of innovative products would be more successful than current approaches.

There are three concepts that I would like to introduce at this point: co-innovation, consumer-led innovation, and forward commitment procurement. I believe that these concepts are essential to develop a sound innovation framework for the food supply chain.

9.6.1. Collaborative Supplier-Manufacturer-Retailer Innovation

In the context of this paper, cooperation is defined as similar or complementary coordinated activities performed by different firms in a business relationship to produce superior outcomes, with expected reciprocity over time. In a cooperative framework, the relationship between partners is driven by the need of profitability and by strategies that are congruent within each company involved and within the relationship.

Co-innovation is one of those strategies, where the success of a new product or service in the market means success for all the supply chain partners. Coinnovation seeks to extend the scale and scope of external partnerships and alliances to access and exploit new technologies, knowledge, and markets. Collaborative innovation requires early involvement of all chain players in the new product development (NPD) and the development of 'innovation networks,' or clusters of collaborators working at every stage of the product cycle and at every link of the value chain. The challenge of co-innovation is better understood if we take into account the balance of market power between food suppliers, manufacturers, and retailers: in the Australian and UK environment, retailers (with emphasis on the supermarkets) are perceived as having more influence and control in the food chain. This influence received significant attention in the UK, which led to the introduction of a Supermarket Code of Practice, introduced in 2002. The Code explicitly mentions that all supply chain participants would benefit if they worked together to expand the market for their products and develop a profitable and sustainable business. In Australia, the ACCC is seeking to investigate the nature of the relationship between suppliers and retailers in a similar manner. Therefore, creating the required level of communication, trust, commitment, and interdependence required for co-innovation calls for significant political and organizational efforts.

9.6.2. Consumer-Led Innovation

Consumer-led product development arguably offers the best platform for co-innovative projects between suppliers, retailers, and manufacturers. The aim of consumer-led product development is to create product differentiation, leading to higher consumer satisfaction, increased levels of consumption of specific products, or increased overall value of the given level of consumption.

In a consumer-orientated approach, the development of new products, processes or services begins with consumer and market research to identify the specific characteristics required by consumers. The next step is the organization of the cooperative framework and information exchange among all the actors in the production chain. An important aspect in this process is the translation of descriptive and qualitative terminology in which consumers express themselves into technological specifications for all the chain players participating in this form of co-innovation.

Consumer-led innovation also opens new doors of collaboration. Consider the highly perishable fruit and vegetables category, which have a shelf life of sometimes days only. The shelf life of the product dictates the speed the supply chain should operate at. Therefore, for these products, some retailers plan their resourcing more than once a day, because the potential wastage cost exceeds the savings through economies of scale in transportation and warehousing activities. In this case, the flow benefits of co-innovation can be captured simply with information sharing and forecasting collaboration. For less perishable items, say, canned food, a highly efficient supply chain depends on low inventory levels and high capacity utilization. In a co-innovative environment, sharing of information has more probabilities of success than in the case of a traditional, manufacturer-led innovation.

Fearne suggests that there are three drivers to undertake co-innovation:

(a) It allows the development of new (value added) products/services for distinct customers and targeted consumer segments.

(b) Process improvement occurs for existing products/services beyond organizational boundaries.

(c) Innovation at the interfaces in the value chain are much more difficult for others to copy, thus increasing the competitive advantage.

Since co-innovation quite often requires global innovation networks, there are further advantages, such as:

(d) Supporting collaboration among geographically dispersed teams of suppliers, manufacturers, and retailers.

(e) Speeding time-to-market and reducing costs for all the parties involved in the NPD process (due to resource pooling and reduction of learning curve).

(f) Meeting customer expectations in a closer manner, due to the direct input of retailers into the main purchase drivers of products.

(g) Enforce consistency and quality of brands and innovations.

(h) Creating a compliance audit trail through sharing of quality documentation.

(i) Most importantly, creating a repository of protected know-how and intellectual property, only available to the chain participants.

How can organizations achieve the right environment for co-innovation? Several authors agree that the following are key aspects to achieve this (Figure 9.7):

(1) Have a consumer focus and enforce collective responsibility;

(2) Promote value chain visibility and information flows (vertical and horizontal) across the chain;

(3) Promote the management of interpersonal and inter-organizational relationship;

(4) Align the required resources (physical, financial, and organizational) with the final demand and process integration.

Product/process	Challenge	Potential need
Food value chains	Identifying improvement opportunities on food value chains	"Increase in 1% the margin profit of tomato growers, without increasing the product price at retail and without sacrificing other margins in the chain"
Fresh, ready-to-eat meals	Developing solutions to overcome the cost–time tradeoffs made by consumers, while also addressing quality and variety demands.	"Develop a range of chilled retail ready-to-eat salads that extend the product shelf-life to 10 days"
Logistics	Finding ways to add value in more than the supply of product	"Develop a logistics system that brings the shop to the consumer, rather than bringing the consumer to the shops"
Supply chain management	1. Manage production in an environment of unreliable climate, including managing the supply and cost of water. 2. Added complexity in food marketing and distribution will require the use of demand signals and fostering relationship management to better understand customer trends and preferences.	1. "Develop horticultural varieties that require x% less water for growth and yet, maintain their desirable nutritional and quality attributes" 2. "Develop a forecasting tool for predicting the monthly demand of highly perishable, ready-to-eat sandwiches in petrol stations".

Figure 9.7: Examples of innovation challenges in the food industry and their translation into market needs.

Source: https://dfat.gov.au/trade/topics/review-of-export-policies-programs/Documents/FoodChainInteeligencePartA.pdf

9.7. A NEW MODEL: SUSTAINABLE CO-INNOVATION

In this section, I use the three concepts discussed in the previous section to propose an innovation model for the food industry. I do not propose that this innovation model supersedes all innovation-related initiatives and it does not offer a solution to maintain curiosity-driven R&D. However, the model proposed tackles the issue of innovation, or R&D with a purpose.

I do believe that this model can replace the fragmented innovation attempts currently undertaken by CRCs and RDCs in the food supply chains. I will use the term "sustainable co-innovation" (SCOI) model to identify the proposed structure.

The SCOI model follows closely the FCP model, adding the concepts of co-innovation and innovation networks. The SCOI model requires a central overseeing organization. This organization (which we will identify in this proposal as "National Food Innovation Strategic Council" (NFISC), can be modeled after the successful NFIS and NFIC. However, we envisage that NFISC would amalgamate the two roles.

NFISC should receive funding from private and public partners (e.g., joint venture). NFISC would:

(a) Provide a strategic framework for food innovation in Australia, from a market-led, supply chain perspective.

(b) Coordinate the activities required to introduce new technology in the marketplace, following an FCP approach:

- Engage with Government departments and food supply chain players to develop the buyers needs into specifications for tenders.
- Act as a technology broker, bringing venture/equity capital and innovator companies together.
- Coordinate legislative activities that provide adequate market conditions for adopting innovations in the wider industry.
- Promote technology demonstrations and case studies to show the business case.

(c) Coordinate government-led food innovation activities (e.g., grants and strategic directions for public R&D), from a supply and value chain perspective. This would avoid the lack of supply chain focus and would increase critical mass in the initiatives undertaken.

(d) Provide timely competitive and technical intelligence to all stakeholders about the particular FCP projects targeted, bringing to the stakeholder's attention current and emerging technologies in the targeted technology markets.

In the SCOI model, a buyer consortium (which may or may not include a public sector organization) commits to purchase a pre-defined quantity of a product\ technology\service, currently under development but not yet available as a commercial offering. NFISC and the buyer consortia agree in the performance sought for the innovation. The supplier consortium is formed, and solutions are sought, based on past innovations or completely new concepts.

Meanwhile, the legislative consortium develops standards, regulations, and certification processes that enable fair competition and enhance the chances of the uptake of new solutions at the agreed performance specification. When the innovation has been developed, meeting all performance criteria, the buyer consortium purchases the product at a specified volume and cost, at levels that encourage other supplier consortium to enter the market. The private sector would react by freeing investment to search for innovations that respond to those specifications. In the context of SCOI, I do not necessarily advocate for the government to become the early market buyer that executes the forward commitment options. This may be effective in some areas, such as food safety (related to public good). However, the Government may not be the ideal buyer innovations that lead to new products, new markets or new processes. Instead, I propose that food supply chain players form alliances that commit to buy the new product/service/technology when this is developed. The drivers for such commitment could be based in:

(a) superior value proposition (including financial, environmental, and social performance);

(b) new regulations encouraging the uptake of the innovation; and

(c) a demonstrated increase of competitiveness in the marketplace if the innovation is adopted.

Selection criteria of publicly funded food innovations

The following general areas that should be assessed for R&D funding are:

1. Health areas

R&D examples:

- Role of human nutrition, genetics, and food trends in health trends (nutrigenomics);
- Functional foods;
- Delivery technologies of healthy ingredients;
- Evaluation of functionality for Australian native foods;
- Portion control and obesity.

2. Food processes

R&D examples:

- Non-thermal processes;
- Micro-machine processing and production-descaling;
- Biotechnology (e.g., development of high-value bioproducts from food industry waste streams);
- Nanotechnology in food production and manufacture.

3. Food supply chains

R&D examples:

- Information and Communication Technologies in modern food supply chains (e.g., track-and-trace systems);

- Food logistics and packaging systems adequate for specific chains (e.g., retail-ready, convenience markets, aging population);
- Quality keeping in perishable food chains;
- Sustainability of food supply chains, with an emphasis on water, energy, and food waste (e.g., carbon-neutral supply chains);
- Improvements in food transportation and storage (cost, quality, and energy);
- Pandemic scenario and contingency planning;
- Tamper-resistant and sanitary packaging;
- Automation in the food industry (including robotics/machine vision systems for production and quality grading);
- Biodegradable packaging;
- Low-cost alternatives for the transport of perishables;
- Role of retail in market-led innovations.

4. Food safety

R&D examples:

- Detection of biological and chemical food contaminants;
- Innovations to counter food bioterrorism threats;
- Authenticity analyses;
- Track-and-trace technologies for food safety;
- Decontamination systems for the food industry.

Although I propose that the SCOI model replaces current funding sources to the food chain (including RDCs and CRCs), I also propose that funding should be available to establish "innovation networks" in key areas affecting the food supply chain environment (e.g., environmental impacts, economics of global supply chains, consumers, and markets, social impacts). The innovation networks should be created in an 'ad hoc' basis and should serve a specific purpose.

For example:

- Brainstorming on particularly challenging issues affecting the Australian food industry (e.g., market barriers, cost effectiveness, labor) and possible innovative solutions.
- Sharing learnings and insights in discussion forums around particular industries (e.g., dairy, meat, horticulture), in a secure environment.

- Providing guidance to food supply chain players, in the form of workshops, seminars, and competitive intelligence reports.

These networks should be dismantled as soon as their goal is accomplished. The lifecycle of these networks is likely to be shorter than current CRCs (e.g., months rather than years) and with input from a larger representation of supply chain players than CRCs structures currently allow.

9.8. CONCLUSION

- One of the main difference between innovation and R&D is that in former, individuals are unaware about the results. For creating an innovation, the required outcome should be known prior to tackling or handling the issue with Research and development.

  All the research owns a purpose, yet a commercializing aim is different from the purpose that seeks knowledge. The funding mechanism for delivering in both these areas are required to follow a unique rationale.

  It is very clear that few of the industries will get the advantages from market push innovation model instead from a technology push innovation model. This food supply chain is considered as an example of a domain that may benefit from the approach of market push.
- There is a need for a concerted approach to innovation in the food industry, from agricultural production and rural-based manufacturing to retail, thus adopting a value chain perspective. Such a concerted approach should recognize the highly dynamic nature of the food manufacturing industry, the benefits of market-driven innovation and the usefulness of policy as an instrument to encourage innovation in the marketplace.
- The current innovation system, which focuses on CRCs, RDCs, public R&D organizations, and universities, has delivered a small number of innovations to the marketplace.
- NFIS had a strategic role in supporting business innovation, manufacturing, supply chain and value addition projects. Additionally, NFIS approached food innovation with a 'whole-of the-chain' perspective. The termination of this initiative leaves a vacuum in R&D focused to food manufacturing and

other areas of knowledge required to introduce food innovations in the marketplace (e.g., supply chain, consumer science, and packaging).

- A new "sustainable co-innovation" (SCOI) model is proposed for the food industry. The model is similar to the Forward Commitment Procurement model, but the concepts of co-innovation and innovation networks have been added.
- The SCOI model requires a central overseeing organization similar to NFIS. However, it is proposed that the organization becomes a joint venture between private and public partners. This organization would:
    a. Provide a strategic framework for food innovation in Australia, from a market-led, supply chain perspective.
    b. Coordinate the activities required to introduce new technology in the marketplace.
    c. Cooperate government-based activities of food innovation (example, strategic direction and grants for public research and development), from a value and supply chain perspective. It may avoid or prevent the lack of focus of supply chain and may increase the mass in the initiatives that are undertaken.

Give timely technical and competitive intelligence to each stakeholder regarding specific areas of innovation being targeted, brining the attention towards present an evolving innovation and technological trends in given areas.

- The SCOI model also requires the following types of consortia:
- A buyer consortium, formed by two or more supply chain partners that establish an alliance to contract/purchase the new process/ product/ service developed by a supplier consortium.
- A supplier consortium, formed by two or more organizations that seek to deliver the innovation at the specifications (e.g., cost, time, performance) set by the legislative consortia.
- A legislative consortium, which aims to increase the receptiveness of the market to new technologies/processes/products by introducing standards, regulations, and laws that increase performance targets in certain areas.
- In the SCOI model, a buyer consortium (which may or may not

include a public sector organization) commits to purchase a pre-defined quantity of a product\ technology\service, currently under development but not yet available as a commercial offering. NFISC and the buyer consortium agree in the performance sought for the innovation. The supplier consortium is formed and solutions are sought, based on past innovations or completely new concepts. Meanwhile, the legislative consortium develops standards, regulations, and certification processes that enable fair competition and enhance the chances of the uptake of new solutions at the agreed performance specification. When the innovation has been developed, meeting all performance criteria, the buyer consortium purchases the product at a specified volume and cost, at levels that encourage other supplier consortium to enter the market. The private sector would react by freeing investment to search for innovations that respond to those specifications.

- In the context of SCOI, I do not necessarily advocate for the government to become the early market buyer that executes the forward commitment options. This may be effective in some areas, such as food safety (related to public good). However, the Government may not be the ideal buyer innovations that lead to new products, new markets or new processes.
- Instead, I propose that food supply chain players form alliances that commit to buy the new product/service/technology when this is developed. The drivers for such commitment could be based in:

(a) superior value proposition (including financial, environmental, and social performance);

(b) new regulations encouraging the uptake of the innovation; and

(c) a demonstrated increase of competitiveness in the marketplace if the innovation is adopted.

REFERENCES

1. Estrada-Flores, S. (2008) *Sustainable Co-Innovation: The Food Supply Chain as a Case Study Submission to the Review Panel of the National Innovation System*. [eBook] Available at: https://dfat.gov.au/trade/topics/review-of-export-policies-programs/Documents/FoodChainInteeligencePartA.pdf [Accessed 16 Mar. 2018].

INDEX

E

F

G

H

I

L

M

N